Praise for *LOVED*

"LOVED *is a supremely practical book and a must-read for anyone wanting to do great product marketing. Martina's stories from the trenches bring everything to life. I'm telling everyone at my company to read it.*"

—Sarah Bernard, CCO Greenhouse, former VP
Product and Design, Jet.com

"*The world is filled with great ideas and products that go nowhere. The difference between nowhere and greatness is product marketing, and Martina is the master. Every tech CEO needs to read this.*"

—Amanda Richardson, CEO CoderPad, former VP
Product, Chief Data, and Strategy Officer, HotelTonight

"*I lived the difference great product marketing makes for a business. Martina's unique expertise on each function in a business makes her advice invaluable.* LOVED *is a book truly worthy of any shelf.*"

—Leyla Seka, COO, Ironclad, former Partner,
Operator Collective

LOVED

The Silicon Valley Product Group Series

INSPIRED: How to Create Tech Products Customers Love, 2nd Edition (Marty Cagan, 2017)

EMPOWERED: Ordinary People, Extraordinary Products (Marty Cagan with Chris Jones, 2021)

LOVED: How to Rethink Marketing for Tech Products (Martina Lauchengco, 2022)

MARTINA LAUCHENGCO

Silicon Valley Product Group

LOVED

HOW TO
RETHINK
MARKETING FOR
TECH PRODUCTS

WILEY

Published by John Wiley & Sons, Inc., Hoboken, New Jersey.
Published simultaneously in Canada.

For general information on our other products and services or for technical
support, please contact our Customer Care Department within the United States at
(800) 762-2974, outside the United States at (317) 572-3993 or fax (317) 572-4002.

Wiley also publishes its books in a variety of electronic formats. Some content that
appears in print may not be available in electronic formats. For more information
about Wiley products, visit our web site at www.wiley.com.

Library of Congress Cataloging-in-Publication Data is Available:
ISBN: 9781119703648 (cloth)
ISBN: 9781119704393 (ePDF)
ISBN: 9781119704362 (ePub)

Author Photo: Courtesy of the Author; © Gary Wagner Photos
Cover design: Paul McCarthy
SKY10032384_021822

To everyone who asked me to recommend a book on product marketing, this is for you.

And to Chris, Anya, and Taryn, thank you for supporting me while I wrote it.

All my royalties from this book are being donated to organizations supporting the advancement of women and underrepresented minorities in tech. Our world is better when tech products are built by the people they serve.

Contents

Foreword

In *INSPIRED*, I argued that the single most important concept in all of product is the concept of product/market fit.

For startups, achieving product/market fit, including and especially the go-to-market strategy for that product, is really the only thing that matters.

But the reward for reaching product/market fit is growth, and growth brings its own challenges.

Moreover, as the company grows, we typically evolve our product to address the needs of additional markets, and usually we soon begin work on new products as well, so the critical concepts of product/market fit and growth remain at the heart of product work for the life of a technology-powered company.

INSPIRED described the techniques we use to discover a product that is valuable, usable, feasible, and viable, and the book discussed how this requires an intense collaboration between product management, product design, and engineering.

But while discovering a winning solution may be necessary, it is not sufficient.

We have all seen countless examples of products that go nowhere because:

- the product doesn't address real customer needs
- or, there aren't enough customers with those needs
- or, those customers do exist, but not enough of them learn your product exists
- or even if they do find you, they don't see how what you provide aligns with their needs

To avoid this fate, as the name implies, there is another dimension to product/market fit, and that is the *market*.

When we talk about a winning product, we are referring to a strong solution for a *specific market*.

A product manager's partner in achieving product/market fit and getting this product to market is the *product marketer*.

While the product manager focuses primarily on the *product* side of the equation, the product marketer focuses primarily on the *market* side, including the go-to-market strategy.

But it's important to realize that pursuing the product and pursuing the market are *not* independent activities. They are happening in parallel, and they are very much intertwined.

Which is why the product manager/product marketer partnership is so important to get right.

I have always visualized this partnership as the product manager working with the product marketer to *triangulate* on product/market fit.

And once we achieve product/market fit, and our focus moves to growth, the collaboration of product and product marketing becomes the key to that growth.

While the product marketing role has existed for many years, for technology-powered products and services, with the pace of innovation and with very crowded competitive landscapes, it is especially challenging, and more important than ever.

At a strong tech-powered product company, the product marketer helps to answer some very fundamental questions essential for a product's eventual success:

- Determining the best ways to reach the target customer
- How and when the customer will be able to learn your product exists
- How to position your product so the customer knows how to think about your product

- How to message the value so that it resonates with the customer's underlying needs
- How the customer can evaluate your product
- Who and how the customer will make a buying decision
- Finally, if you've done your job well and the customer loves your product, how they can tell their friends and colleagues how much they love your product

Many experienced product leaders will tell you that getting the go-to-market right is as tough as discovering a successful product.

In truth, in our books and articles to date, we know we have focused primarily on the product side of the equation.

That's mainly a result of our product bias. We know there are examples of products that succeeded despite weak product marketing, but great product marketing can't overcome a bad product.

However, in our increasingly competitive reality, in order to succeed, we need *both* strong products and strong product marketing to succeed.

Which is why I'm happy to tell you about this new book.

Martina has had a remarkable career, with many years of experience at top tech companies, most notably Microsoft and Netscape Communications, covering not just product marketing, but also product management and corporate marketing. She is, I believe, uniquely suited to write this book.

Martina has worked for, and been coached by, several of our industry's most accomplished technology and marketing leaders. As a long-time SVPG Partner, venture capitalist, and UC Berkeley Lecturer, she has been advising, coaching, and teaching literally hundreds of companies and countless product marketers on the critical topic of product marketing.

In some cases, especially at early-stage startups, the product manager may need to cover the product marketing role as well.

In other cases, others in the marketing organization may need to cover the role.

Whether you are coming from the product side or the marketing side, you are much more likely to succeed if you have a solid understanding of product marketing.

It is the goal of the Silicon Valley Product Group series of books to share the best practices of the top product companies, and this is an important addition, addressing a long-underserved need.

And our intention is that this is just the start. We plan to do more going forward, sharing more of the best practices and techniques that help product teams and product marketing to collaborate effectively and successfully.

Marty Cagan
November 2021

Introduction: My Story

Getting Flamed by Bill Gates

When Blue walked through my door, I knew it couldn't be good. The only other time the Word Business Unit manager sat down in my office was back when he was doing his whistle-stop get-to-know-you tour. He got right to the point.

"I just got an email from Bill Gates. It said, 'Word for Mac is depressing Microsoft's stock price. Fix it.' So, I'm here to ask, what are we doing?"

I was a young product manager for Word for the Mac, and it was the first time I'd been trusted with a major product. A few months earlier, the newest version of Word for Windows released, delivering against a strategic plan that was years in the making. Up to that point, the Windows and Mac versions had different code bases, features, and release cycles. This new version used a single code base for both, meaning for the first time, the two would have the same features and ship simultaneously.

But the Mac version was late—very late. Each day it slipped past the Windows release was viewed as a public failure. We rushed to get the product done, deciding its new features were worth a hit in the product's performance.

Mac users HATED it. It was so slow that in their eyes it felt barely usable. And they missed their more *Mac*-centric features.

Back then, Word and Excel were the most significant productivity products on the Mac. Apple was a beleaguered company, and if Word didn't work well, there was real fear in the Mac community that it could be the death knell of Apple.

Newsgroups spewed vitriolic hate at Microsoft. When I posted to earnestly defend our decisions, they directed that

1

hatefulness at me. I would sometimes end my days in tears, wondering, "Don't people realize I'm a person?"

The only way to "fix it" was to improve performance and the features Mac users cared about most. We released a significant update along with a discount voucher and a letter from me apologizing to every registered Mac user.

It was a humbling experience. But it taught me an important lesson: the market determines the value of a strategy. And even at a company as good at strategy as Microsoft, things can still go really wrong when a product goes to market.

Start with the End in Mind

Although it didn't get everything right all the time, Microsoft did do a lot of things right much of the time. Working there was like going to the university of software because you got to see so many products succeed and fail in so many different markets. In every case, Microsoft lived by the disciplined application of objectives, strategy, and tactics, always starting with the end in mind. Beginning my career there deeply shaped me.

Every move, even small ones, mapped to Microsoft's strategic objectives. I arrived at Microsoft just as it was preparing to launch the first integrated version of the now ubiquitous Microsoft Office suite. In all our product collateral, we removed mentions of "desktop productivity applications"—the old category name—and instead used "integrated office suites." Ever playing the long game, it was part of shifting the category, reinforcing the notion that Office was the standard-bearer.

I watched how a systematic approach of combining great products with equally great market strategy killed our two biggest competitors at the time: WordPerfect for word processing and Lotus 1-2-3 for spreadsheets. Just a few years prior, these best-of-breed competitors seemed untouchable. Their failure lay in focusing on features versus building and marketing a bigger vision.

I was a product manager on the Office team when a small pocket of the industry was starting to focus on a relatively new thing called the World Wide Web.

The company that was changing the game, however, wasn't Microsoft. It was Netscape, the originator of the commercial Internet browser. Its threat was so significant, Bill Gates sent an email to the entire company saying no other competitors mattered right now.

That email came just after I had accepted a job as a product manager at Netscape. Understandably, I was asked to pack up and leave the Microsoft campus immediately.

The Barefoot Guy on the Cover of *Time*

My parents could not comprehend why I would leave the storied Microsoft to join a company whose founder, Marc Andreessen, was featured on the cover of *Time* magazine barefoot, sitting on a gold-gilded throne.

I arrived expecting an equally strategic adversary to Microsoft, one playing chess a few moves ahead. But if Microsoft was the command-and-control–style dad, then Netscape was the laissez-faire, chain-smoking uncle. New products or programs were cooked up overnight and announced in a press release. Teams scrambled to make them a reality. There was no formal launch process or standard way of doing anything. It was complete and total culture shock.

But it was where I first experienced the foundation of how modern product teams operate. I bounced back and forth between leading product management and product marketing teams. It let me work with many different empowered engineers who were allowed to experiment and innovate.

Traditional go-to-market was bypassed, distributing over the Internet directly to customers—a completely novel concept at the time. "Products" had public, not closed, betas—again, a totally new idea back then—and were released with minimum functionality that met just enough market demand to create early evangelism as much as they crowd-sourced quality.

Despite all I knew about the value of strategy, Netscape was where I learned that free-range discovery could inspire innovation no one could foresee, at equally unforeseen market velocity. It was a much more dynamic model of company building with higher highs and lower lows.

It was also where I saw how innovative ideas can give birth to new startups.

Markets Shape Success

Ben Horowitz was the most revered executive at Netscape when he chose to co-found a then-new startup called Loudcloud (later Opsware) with Marc Andreessen, Tim Howes, and Insik Rhee. It was the world's first Internet infrastructure-as-a-service company long before the world had a framework to understand it.

Back in 1999, it was a radically new idea and not at all clear that by 2021 95% of Internet data center load would be for cloud traffic.[1] While the vision was there, the services required and architecture of Internet infrastructure at the time—no matter how much our software automated—was too expensive to deliver in a cost-effective way.

I got schooled on the limits of company and category creation while leading marketing and being Ben's chief of staff. I learned my own professional limitations, facing the pain of what felt like failure (more on that later). I also learned that the greatest minds, vision, and plans aren't enough if all the right market elements aren't in place.

---

[1] https://newsroom.cisco.com/press-release-content?type=webcontent& articleId=1908858.

How to Use This Book

In my post-Loudcloud years, I started doing product market-ing advising. I taught workshops for companies like Google and Atlassian and created a class on marketing and product management for engineering grad students at UC Berkeley. I practiced product marketing daily with early-stage startups at Costanoa Ventures and watched startups get acquired and IPO. I observed product marketing in action across hundreds of companies.

Through it all, I learned this: There is a stark contrast between how *most* companies do product marketing and how the *best* companies do it. It's largely because product marketing is misunderstood; it is *the* most foundational work required to market any tech product.

That's right: what you want most from marketing—a bigger pipeline, a loved brand—isn't just about doing more market-ing, it's about doing *better product marketing.*

This book is an invitation to rethink tech marketing by understanding how much product marketing shapes the foundation on which the rest of marketing builds.

You'll need great people to do the job eventually, but strong product marketing can actually be done by whomever has the capability and mindset. It's why I wrote this book for anyone with product or marketing in their purview regardless of title.

In Part 1, you'll learn how a Midwestern code slinger beat a Silicon Valley icon by applying the fundamentals of product marketing. You'll then see each in action as I explain them in depth.

Part 2 explores the people and process parts. You'll learn the ideal profile for product marketers and how they partner best with other functions. I'll also cover crucial tasks and techniques—like how to discover market fit—important to succeed in the job.

Parts 3 and 4 go in-depth on the strategy and positioning work that's so critical and hard to do well. The tools I introduce in those sections have been used with every size and stage company, and they consistently provide a framework for improvement.

Part 5 focuses on the leadership and organizational challenges of product marketing: how to lead it, hire it, guide it, and adjust its purpose at different company stages and business inflection points.

There is one big assumption in everything I write: you can't succeed in go-to-market without a strong product. If you're not yet there, please read Marty Cagan's *INSPIRED*. It focuses on how to build products people love.

Then, when you're ready for your product to be loved by your market, read on.

Part One

The Foundation: Understanding Product Marketing's Fundamentals

Chapter 1

When David Beats Goliath

Why Product Marketing Matters

Marco Arment had the Silicon Valley "It" factor. A prolific developer, he was the lead engineer and chief technology officer of Tumblr, a microblogging website that was sold to Yahoo for over $1 billion in cash. His blog was viewed more than 500,000 times a month, and he had a popular podcast before podcasts were a thing.

It's no wonder tech press were captivated by Instapaper, Marco's next creation after Tumblr. They talked about the app for saving web pages to read later as if it was the only one that did the job.

But around that same time, Nate Weiner, a self-taught code slinger from the Midwest, had seen the same problem. People saw articles in their social feeds or web pages and wanted to save and view them later. He created Read It Later to do just that.

With a dash of visual design from his girlfriend and some coding help from his twin, in just a couple of years, Read It Later was used by 3.5 million people—nearly triple the users of Instapaper—and had hundreds of rave reviews. Yet press talking about great productivity apps still only mentioned Instapaper.

Apple announced a feature called "Reading List" at its World Wide Developer Conference. It validated Nate's app but also prompted a brief Twitter storm, with some declaring game over for Read It Later.

Instapaper stayed its course, occasionally adding new features. Three years after he created it, Marco sold Instapaper to Betaworks. Growth languished. Eventually, what remained of the company bounced around in a game of musical owners.

In that same period of time, Read It Later rebranded as Pocket, won nearly every major app award, integrated into hundreds of apps, got multiple rounds of venture capital, and nailed every external marker of success. By the time Pocket was acquired by Mozilla, the makers of the Firefox browser, it had 20 million users.

How did Nate and his small team beat the Goliath reputation of Marco and Instapaper to win their category?

Despite none of them having the title "product marketer," they collectively worked to shift focus from just building the product to a "product marketing" mindset.

This is just some of what they did:

Sharing data around shifting trends in consumer behavior.
From company blog posts with surprising factoids—for instance, of the thousand most saved videos, the median length was 30 minutes—to showing the press the rapid growth in saving items as mobile device use exploded, the team promoted a customer and market-centered point of view, not just one about their product.

Connecting their product's purpose with broader trends.
They started comparing what they did for web pages to similar shifts. Like how Dropbox changed file sharing or Netflix changed TV. They connected themselves to a much bigger "anytime, anywhere" megatrend, saying "we're the ones who are doing this for Internet content." They also developed an API that let any app integrate their "save-for-later" functionality, making it an industry standard.

Rebranding from Read It Later to Pocket. This was a strategic decision designed to help the world see Pocket as bigger than saving articles for later. Changing the product name when they released their 4.0 version was important to show some of their key differentiators—like the ability to save videos and images—and define what was important for products like theirs.

Making it free instead of $3.99. It's hard to ask people to pay for something when they haven't yet experienced its value. The change was messaged in an authentic blog from its founder, where he also explained they were now a venture-backed startup. It helped cement the trusted relationship Nate had with Pocket's users despite big changes.

Sharing the "why" and advance access with influencers. Before launching any new version, they made sure to give the most influential evangelists—press, pundits, superfans—the "why" behind the new enhancements, giving new features more meaning.

Like many who build products, Nate's initial instinct to beat Instapaper was to add more features and be the better product. While product enhancements were critically important (they did a major redesign along with the rebrand), without a market context that gave them meaning, it would have just been more feature noise in a world already drowning in apps. The natural inertia behind Instapaper would have kept it the industry's darling.

Pocket's story is like many others in tech. Competitors are bigger or better known. The product team is concerned the world hasn't heard of them nor do prospective customers understand what they've built or why it matters. The impulse is to build more product to show why it's better. And while you have to build good product, market traction—how every product's success is ultimately measured—requires equal, concerted effort on the market side. Specifically, who's the right market, the best ways to reach them, and who needs to say or do what for your product to be credible.

This is the job of product marketing.

What Is Product Marketing?

Product marketing's purpose is to drive product adoption by shaping market perception through strategic marketing activities that meet business goals.

The work is not optional. As the Pocket team discovered, if you don't position your product and act with clear purpose, competitors and market dynamics actively work against you.

Product marketing brings strategic intent and product insight to all market-facing activities. It coordinates a winning plan across the entire go-to-market engine (marketing and sales) and provides the foundational work for everything those teams need to succeed. It's necessary work for everything from hitting user goals to leading a category. If you look at the list of what Pocket did, everything framed why Pocket had value even when they weren't talking about their product.

The job also includes working with product teams to make better decisions affecting market adoption. This can range from prioritizing a feature to writing a piece of content that reframes a competitor. When Pocket wrote blogs about top saved videos, it highlighted saving videos—a feature Instapaper didn't have—and established what good products in the category do without talking about their product at all.

The work is both highly strategic and tactical. But it is far more than creating product collateral, doing sales enablement, or managing launches, a common misperception of the role. Unfortunately, for too many, this collection of tasks has become the role. These tasks are a function of the job being done; they don't define its purpose.

Fundamentals

My hope in writing this book is to refocus product marketing around its purpose—leveraging product investments in a deliberate way so the go-to-market machinery can achieve a business's goals.

That requires clarifying what it means to do the job *well*.

It starts with the foundation of product marketing, which comes down to just four fundamentals from which all important work flows. They are:

Fundamental 1: Ambassador: Connect Customer and Market Insights

Fundamental 2: Strategist: Direct Your Product's Go-to-Market

Fundamental 3: Storyteller: Shape How the World Thinks About Your Product

Fundamental 4: Evangelist: Enable Others to Tell the Story

The rest of part 1 of this book explains these fundamentals in depth and how to do them better.

Why Product Marketing Matters Now

I'm intentionally using the function **product marketing** and not the role **product marketer** as I describe all this. As Nate and the team at Pocket showed, good product marketing can happen if capable individuals are willing to do the work. Nate and his team were exceptional at learning, and their willingness to apply product marketing fundamentals was outstanding.

Not every company has people willing or capable of doing this, but it does mean you can get product marketing done without the perfect team formation yet in place. This doesn't mean you shouldn't staff the role. Outcomes always improve with strong product marketers. It just removes excuses for not doing the work with who you have now.

Because the need for good product marketing has never been more urgent or important.

Modern development tools (open source, cloud every-thing) and methods mean every product landscape isn't just increasing. It's increasing at an increasing rate. For example, the marketing technology category had 150 companies in its first year and over 8,000 battling it out nine years later. The Apple App Store opened with 500 apps and now has over 5 million. There's the API economy, Web 3.0, the rise of product-led growth. Search engines and tech giants are the front door to customers discovering nearly all product information. Social media doesn't just influence points of view, it contains millions of influencers—at least 100× the number of journalists in the world.

Products make identical claims and have similar features. Pricing often doesn't help frame value—similar products can have very different prices for non-obvious reasons. Trusted rela-tionships and word of mouth are more powerful than ever in influencing decision-making—even for major enterprise soft-ware. Imagine how difficult it is for any potential customer to navigate the modern decision landscape.

There is no way for a product to stand out and win unless its entire go-to-market engine is carefully coordinated and it holds a clear market position. That's product marketing.

Where Product Marketing Fits

There is a lot of confusion about the difference between marketing—the function at large—and product marketing, the specific role often part of the marketing organization.

The customer journey is never a straight line: They expe-rience a problem, plunge into the information landscape, and may at some point come out with a hand up to try, buy, or enter a sales process.

The art of finding and reaching that customer on their jour-ney with the right message at the right time so they are willing to consider a product is the job of *marketing*. The art of selling and converting a prospective buyer into a customer is the job of *sales*.

Modern marketing teams are filled with specialists who amplify messages, deploy campaigns, and manage and execute programs in their respective domains: demand generation, digital/web/search, advertising, social, content, influencer, community, analyst relations, marketing operations, public relations, marketing communications, brand, and events—many residing in a corporate marketing function. The range of marketing channels is so vast, I have a dedicated appendix at the back of the book defining them. The bigger the company, the more complex and layered marketing becomes.

Marketing specialists rely on product marketers to do their jobs well. Product marketers define what aspects of a product to promote, who to target, why target customers care, and which channels are most important. They are the bridge between the product organization and ensuring the actions from the go-to-market engine of marketing and sales result in business impact.

Part 2 of this book dives into the details of how product marketing interfaces with all its partner functions—product, marketing, and sales—and the best practices that make these partnerships effective. The rest of part 1 focuses on explaining the four fundamentals in more detail.

Practicing them is what gives clarity of purpose to a product's go-to-market. That intentionality, in turn, is what separates companies that succeed wildly from those that do just well enough. It requires investing in a strong product marketing foundation, and the purpose of this book is to show you how to do it.

Chapter 2

The Fundamentals of Product Marketing

You know that red squiggly underline that automatically appears any time you've misspelled something in Word?

It debuted after an executive team edict proclaimed every product unit had to simultaneously ship its next version together with the upcoming release of Windows. The team's development time was slashed in half, which meant we'd only be able to ship a fraction of the number of features as the prior version.

This was in the peak of the feature arms race era, a time when value was equated with stickers on boxes highlighting hundreds of features packed inside.

Not only were there far fewer features in the next version, but many of them weren't major game changers. They were clever enhancements of features already in the product. How could we take our feature-light version and make it feel like a worthy, full-fledged release?

It was in a team brainstorming session that a product marketer pulled out an instrumentation study the product team had done. It analyzed every keystroke of hundreds of users. He pointed out the planned enhancements fell into two categories:

1. Functions most people used most of the time—like formatting text

2. Features used less frequently, like bulleted lists, but for the people who did use them, they used them a lot

It was a eureka moment—this is how we can tell this version's story: *it focuses on what matters most for how most people use Word.*

Back then, press and analyst meetings mattered a lot because one review could define a product's reputation for years. Product marketers would go on dedicated in-person tours showcasing the product to key influencers and pundits.

Usually, these meetings led with a polished PowerPoint. The product marketing team decided to ditch the slides and instead freestyle on a whiteboard (Figure 2.1) to share the data and have a conversation. They then demoed the product using a story of how an ordinary office worker used Word in her daily work.

The story that went along with it went something like this: *75% of the actions someone takes in Word fall into basic categories like formatting and file management. We focused our features on those areas*

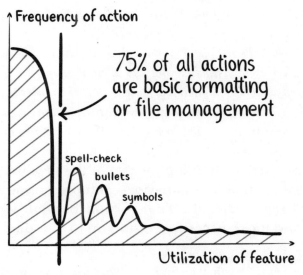

Figure 2.1 Whiteboard recreation.

so every user could benefit from them. When we examined how features actually get used, we also saw some that aren't used by as many people but for those who do use them, they use them a lot. That told us those features have a lot of value but aren't easily discovered. This version implements those features on behalf of users so they can experience their value without changing how they work.

Case in point: spell-check. It now ran in the background and underlined misspelled words as people typed instead of making someone remember to hit the spell check button.

Walt Mossberg, then of the *Wall Street Journal* and the most preeminent word-processor reviewer of the time, had a justly earned reputation as immutable. When he asked why this version had a highlighter feature, beyond showing him the research, we sent him an email from a beta user raving about the feature and how much she used it.

When his review finally came out, it surprised even us.

Welcome to this week's colum. Wait, that's wrong. I meant this week's column. There, that's better. The word-processing software with which I'm writing this noted that typo and flagged it for me by instantly underlining it with a squiggly red line ... For the new Word, also known as version 7.0, Microsoft has concentrated on a host of small but clever refinements that automate and enhance the writing process, like the aforementioned "Spell-It" and improved "AutoCorrect" features. It converts the asterisks or hyphens you use to mark items in bulleted lists into indented symbols ... There's a feature that simulates a yellow highlighter pen ... Taken together, these new features make an already excellent word processor even better, in my view. Word continues to be the best writing tool out there. [1]

The approach of framing the *why* of the product's features together with actual user behavior—and presenting it in a less traditional way—paid off. That version became Word's best reviewed and most successful version up to that time.

---

[1] Walter Mossberg, "Personal technology: Word for Windows 95 Helps Sloppy Writers Polish Their Prose," *Wall Street Journal*, October 5, 1995.

Product marketing worked with the product team to come up with compelling positioning and spearheaded everything else to bring Word to market: coordinating a product launch, creating sales tools for the field, readying customer testimonials, influencing pricing, enabling product evaluations, preparing competitive response tools, educating channel partners, and working with direct marketing and advertising teams to create compelling campaigns.

These are tools of product marketing's trade but remember, they aren't its purpose. The job is to drive product adoption by shaping market perception through strategic marketing activities that positively impact the business.

The work of Word's product marketing team made the difference between the product just going to market versus setting the standard for the category. It did this with the focused application of the four fundamentals in everything it did. They were **ambassadors** between customer, product, and market insights that enabled great **storytelling** that positioned the product. This set the foundation for activities and tools that let others **evangelize** through all the go-to-market machinery. This was all guided by clear **strategies**, the most important being to leverage a new operating system release.

Let's explore each of these fundamentals to explain the range of activities they encompass and how to do them well.

Fundamental 1. Ambassador: Connect Customer and Market Insights

Everything a product marketer does must be grounded in customer and market insights, which is why it is the first fundamental. Product marketing brings this expertise to how a product makes its way to market. This goes beyond knowing what problems a product solves and who target customers might be. It's about providing market and customer insight into any situation.

The range of work encompasses segmenting customers, knowing frustrations and problems that drive them to seek something new, and delineating the steps they take along their journey to become customers. It includes knowing what makes raving, loyal fans and the "watering holes" that create influence or amplify influencers. It means understanding which activities engage future customers as well as existing ones.

It also requires deep product knowledge of what customers find useful. Product teams should already have a clear idea of why customers choose to use or buy a product. Product marketing brings: knowledge of the buyer's frame of mind, how the competitive landscape might affect a decision, and what that means for how a product should be positioned.

It is both quantitative and qualitative knowledge, a shared effort with product, insights, or research teams. It's about understanding how desired customers think and act, then applying it to product go-to-market.

Fundamental 2. Strategist: Direct Your Product's Go-to-Market

How do you make every market-facing activity you do matter? By defining a clear product go-to-market plan whose strategies align directly with business goals.

Strategies guide the tactics (activities) that get you from A to B. A good product go-to-market makes clear the *why* and *when* for certain activities in addition to the *what* and *how*.

Product marketing considers why a customer might want a product and how they are likely to find it, then plans accordingly. Does a customer rely on peers at networking events to learn what's new? Does she primarily do her research online and prefer to try new technology herself?

Much like product managers use discovery techniques to determine if a product is valuable, usable, feasible, and viable, in product marketing, aspects of product go-to-market can only be discovered by trying things in-market. For example, a company might not know whether or not trial or product-led growth will be an effective way to grow versus using a direct sales force. These are expensive investments, but the range of go-to-market models are now broad enough that which is best for a business can't be presumed without some experimentation.

This means defining a strong product go-to-market is iterative. The purpose behind activities is made clear, then thoughtful actions are executed—both planned and opportunistic. Learnings are then applied to evolve a product's go-to-market plan. It's normal that some things don't work well. A product's go-to-market plan must be resilient enough for experimentation and failure.

That's why defining strategies—and all their associated activities—well requires a strategic *and* learning mindset.

Fundamental 3. Storyteller: Shape How the World Thinks About Your Product

Not everything that's said about a product is in a company's direct control. But the foundational positioning work that strongly shapes how the world thinks about a product is.

Positioning is the place a product occupies in people's minds. It sets a context in which a product's value becomes clear. The key messages supporting that positioning are what marketing and sales teams say and promote over and over to reinforce that position. A broader story narrative is what stiches it all together to make it stick.

Positioning is a long-term game, while messaging is a shorter-term one. But success with both requires a combination of perseverance (as you find what works) and patience (as you build on it).

For positioning, every marketing action can reinforce position, but defining the goalposts by which your product and category are measured are some of the most important. For messaging, the range of work includes iterative development of messages that connect with the right audiences and helping people make more informed decisions.

In the modern era, this means being genuinely helpful without being overly promotional or authoritative. It requires restraint and knowing what is most meaningful to audiences, not trying to say everything all the time.

Our brains process stories differently than straight facts. It's why positioning and messaging is best done through stories—an employee's, a customer's, the product's, or the company's—and why product marketing must be good at telling stories that bring it all together.

Fundamental 4. Evangelist: Enable Others to Tell the Story

Another benefit of stories is they are easier for others to tell. Today's hyper-competitive environment means you can't expect to sell a product that doesn't have other people talking about it.

Evangelism only works if it feels authentic. The range of work to make this happen includes providing direct sales teams with the right messages and tools to make them sound like genuine advocates, not just salespeople trying to sell.

Evangelism also means finding the most meaningful influencers that move your market—key customers, analysts, pundits, press, bloggers, social influencers, online forums—and inspiring them with stories and evidence so they advocate for your product. In the broad digital landscape you see this in reviews, press articles, analyst reports, communities, or any social or developer communication platform, and includes all in-person events and evangelism too.

Healthy growth of any business relies on this organic fly-wheel working well. It's the only way for organizations to scale their market footprint cost-effectively.

The next four chapters explore each fundamental in more depth with stories that show you what they look like when applied and techniques to help you do them well.

Chapter 3

Ambassador

Connect Customer and Market Insights

Whan Julie Herendeen was VP of Global Marketing at Dropbox, her team thought they knew their customers. With users in the tens of millions and all the data that came with them, they felt confident customers were divided broadly into two categories—consumers who acted like microbusinesses and larger companies that had more enterprise-style needs—and marketed accordingly.

She decided it was important for her entire team—not just the product marketers—to get out of the building, pack their bags, and visit customers at their home offices or office parks. Similar to the *jobs-to-be-done* framework, she had her team focus on what customers were trying to accomplish and what was motivating the why behind their choices.

Julie immediately got calls from her team saying, "This is amazing. I'm learning so much" and "I couldn't see any of this in the data." When they got back and crunched through their learnings, they realized some of their assumptions missed the mark on why customers valued Dropbox.

Yes, they were smallish businesses, but they needed to easily collaborate on big jobs—like sharing daily video shoots with a client on a commercial production—and Dropbox gave them the way to do it.

The customer visits also revealed important aspects of how Dropbox customers liked to feel. They valued the freedom to work with whomever they wanted, however they wanted. These much more nuanced insights made clear how her team needed to market differently. They shifted messaging and marketing channels and created all new advertising campaigns.

The experience of Julie's team is precisely why connecting customer and market insights is Fundamental 1 of product marketing. While it is often funneled through product marketing to other marketing functions, in Julie's case, her entire marketing team benefited from their in-depth experience understanding customers.

Most underestimate how nuanced and layered both customers and markets are today, and just how much time and work it takes to truly understand them.

Market Sensing

Markets and the customers swirling in them are never a monolith. Yet they often get generalized into broad categories— like *small businesses*. Modern go-to-market requires understanding nuances around not just what customers are trying to do, but their entire journey toward product consideration— like which products they already use and compare a product to.

At a minimum, here are some baseline product marketing practices to stay connected with customer and market realities:

- **Have direct customer interaction**—ideally weekly.
- **Develop a standard set of open-ended questions** to ask customers or prospects.
- **Reflect insights into product and go-to-market** team discussions.
- **Write the most important insights down** so they can be easily shared and used.

Because every market is crowded—sometimes with thousands of companies in adjacent spaces—it makes defining the precise customer or their journey really challenging. To understand what people *actually* do, use, or value, you must test market assumptions in real life situations.

This is clarified through customer discovery work. Think of it as the market side of product market fit. It must be probed right from the start just as product is being explored. Market fit work is not the sole purview of product marketing. Everyone in a product team (product managers, designers, researchers in some organizations) and the go-to-market engine (marketing and sales) can do the work and shape what is learned.

But not all customer insights are equal in unlocking markets. Product marketing *is* responsible for deciding which key learnings help go-to-market and product teams do their jobs better. Will an insight help a team make a decision or tradeoff on what to say or do next? If the answer is yes, it's additive. If the answer is no, archive it. Strong product marketers help teams stay focused on what matters most.

Product marketing should try to answer market-sensing questions and understand their implications **across the entire buyer's journey**, including both rational and emotional motivations:

- What are they trying to do?
- Do they recognize and prioritize this problem?
- What is motivating them to solve the problem?
- What compels them to take action?
- What in this product delivers the most value?
- Who is most likely to value and buy this product?
- What starts the journey toward acquiring the product?
- How might a product get discovered and become more desired over the entire journey?
- How might we reduce friction in acquiring the product?

- What do people need to see or hear to become customers?
- How can we delight customers so much that they want to talk about the product with others?

Although the answers inform every aspect of product go-to-market, rarely are they clear or complete at the start. As with products, learning the market side of product fit is a dynamic process. Start with a reasonable hypothesis and use everything in market—websites, emails, sales conversations—to iterate toward answers. Adapt based on what is learned.

Chapter 11 goes into more depth on specific techniques to deepen customer insight, such as customer interviews, sales call shadowing, or leveraging any of the ever-growing number of tools in the marketing, sales enablement, or product analytics space.

Third-Party Insights

A market is shaped by more than what teams observe directly. It is influenced heavily by the ecosystem surrounding it.

This requires developing a regular practice of gaining insight from third-party data, research, reports, articles, websites, reviews, press, and social media. Third-party content has the benefit of revealing competitive insights and is a great way to tune into public perception.

Google search trends over time show the natural tendencies of how one set of words compares to another to discover something. When trying to assess topics audiences engage in more deeply, check one of many content services that show the top-read content on a subject.

If you're in a more mature company, you might be lucky enough to have a dedicated customer insights, customer research, or data analytics team. They are a meaningful shortcut in gaining market and customer learning. Use them!

A product marketer's job is to intersect what is learned from direct customer feedback with what is learned from third parties and use it to inform internal conversations. Then, direct the market approach accordingly.

Remember at the start of Pocket's journey when it was Read It Later, despite having many more users and supported platforms it was still not thought of as the category leader? Their product marketing challenge was to move the ecosystem's perception toward their realities; it was part of a bigger shift in behavior and used by more people. It is very different work than needing to convince users a product has value. Understanding and directing activities according to the market challenge based on customer and market insights is what product marketing does.

The Competition

The role of the competitive landscape can never be underestimated. The impact a competitor can have on changing market circumstances often takes companies by surprise. Take these real-life examples:

- A competitor—with no changes in product—adapted their sales process and started winning more deals in head-to-head competition.
- A competitor hadn't even released a product but wrote about their point of view so well and often, if people searched for a solution, this competitor's points of view topped all search results. They were perceived as the market leader before they even shipped a product.
- A company in an adjacent category put on an event that got press coverage promoting features the category-leading company didn't have. It left them scrambling to respond with their sales team and in public forums, making them look like they were playing catch-up, even though they were the category leader.

In each of these examples, the company's product was unchanged. Yet the market reality very much had. While you can't let the competition dictate your course of action, you can't ignore how much they can shape *perceptions*.

That said, beware of over-rotating toward competitive response. Companies can lose their own way if overly responsive to competitors' agendas rather than staying focused on what's best for their customers or market. This is a place where product marketing response can and should be much more dynamic than product's.

Meet challenge with challenge. Outplay the competition whenever you can. But think of this as chess; move ahead of your competition, don't just respond.

Product marketing helps the company stay the course for what is most important and exercise judgment on what merits response.

Ambassador of Insights

Product marketing is the ambassador for all these customer and market insights, so they must be a part of the right internal conversations. They can accelerate a product feature or make sure an engineering blog gets written to diminish a competitor's approach. Product marketing directs appropriate response, whether it's through product, marketing, or sales.

Customer insights can sometimes be heard as customer requirements. These are two distinct things. When product marketers bring customer or market insights to a product team, it's important to see them for what they are—a way for the team to make smarter decisions based on market realities, not just technical ones. How any insight impacts product priorities falls to the product manager to decide.

Customer insights also often get translated into artifacts—like Jobs to Be Done (JTBD) stories (often used by product), Personas (often used by design or product), Ideal Customer

Profile (ICP, often used by sales), and customer segmentations (often used by marketing). Each has a purpose that is specific to the function.

For example, an ICP might have overlap with problems the customer is trying to solve in a JTBD story. But the ICP is used to determine account fit and likelihood to buy, a function of the technologies already in use, the size of the organization, budget availability, and the presence of an internal champion—none of which appears in a JTBD story.

As the ambassador for customer and market realities, the product marketer ensures the most important attributes *that drive customer and market adoption* are known and documented so teams can do their jobs better.

Market and customer insights can add gas to a burning marketing fire. Anything in customers' daily lives, news, or trends can be a propellant if the right moment is seized.

That's why a deep understanding of customers and markets is Fundamental 1 of product marketing. It is simply the foundation on which everything in a product's go-to-market gets built.

Chapter 4

Strategist

Direct Your Product's Go-to-Market

The launch of Pocket 4.0 set the small company on a totally different trajectory. Nate, now CEO of a venture-backed startup, was up at the white board as the team gathered to solve the problem of how to follow 4.0. His marker squeaked as he wrote 5.0's key features on the wall-sized canvas. The affable CTO stared at the whiteboard and asked, "How do we make all this matter?"

Unlike B2B companies, where an analyst can anoint a company "leader" and shape its future, the fate of a consumer app like Pocket's lies in the hands of ordinary people. An app can have a moment but then fade. The team needed a way to boost and grow people's interest. And they needed something press-worthy to compete with hundreds of thousands of other apps vying for attention.

Crowded into the sole conference room with a window, Nate wrote the marketing strategies guiding Pocket on the whiteboard: grow a loyal user base, define and lead the category, and leverage partnerships for growth. Using them as guides to judge each idea, the team agreed an online launch like 4.0's wasn't enough. It wouldn't define the category nor elevate Pocket's importance to potential partners. They needed something that let them tell the whole story of why saving Internet content to view later enabled mobile lifestyles and was good for content creators.

The solution came in an idea internally called Pocket Matters. It was an in-person 5.0 launch event with press, partners, and 10 Pocket users at a San Francisco wine bar. Nate gave a presentation that told the story: long-form content had an avid audience when you let people save and view it later—which is why Pocket mattered—and then introduced 5.0's highlights to the world. They shared a digital media kit for journalists summarizing everything being announced. Partners, customers, and press mingled with one another before and after the event.

Within hours of the event, they had a flurry of press, downloads, and partner discussions that accelerated. Less than a month after the event, Nate was also named one of Time Magazine's 30 People Under 30 Changing the World.

While the event helped with all of their strategies, its purpose was to define and lead the category. This clarity helped them make smart go-to-market decisions *with better results*. It's why being a strategist that directs a thoughtful approach to a product's go-to-market is Fundamental 2 of product marketing.

Key Terms

Throughout this book, I refer to a set of concepts around go-to-market and strategy that, out in the real world, are used loosely. For the purpose of this book, I'll now explain what *I mean* whenever I use terms and how they interrelate. I'll also mention how others refer to them to help clear up what I know can be confusing.

- **Go-to-Market (GTM) Engine**, aka marketing and sales, GTM strategy. This is the sum total of *all* the marketing and sales machinery that bring products to market.

  At scale, it's an engine that picks and chooses how it leverages products. Because marketing and sales activities exist outside any individual product's go-to-market, the term go-to-market can't be presumed to be associated with a particular product. In this book, I'll refer to this as the *GTM engine* (my term) because it crosses functions and organizations and to avoid confusion with other uses of GTM.

- **Marketing Strategy**, aka GTM Strategy. This drives the orchestration of the *marketing* elements in the GTM engine, for example, brand, corporate communications, demand generation, or promotional programs. This is owned by the marketing team at large at the company level.

  At the level of an individual product, marketing strategies are driven by the product marketer to create alignment in a *product go-to-market plan*, where specific activities, how they get done, and dates come together. For most one-product companies, marketing strategies and a product's marketing strategies are largely one and the same.

- **Product Go-to-Market.** If you're familiar with other SVPG work, go-to-market means for a particular product. But since this book places that work into a company's larger go-to-market context, I will refer to a *product go-to-market*—the unique purview of product marketing— when I mean the path for how a particular product goes to market.

- **Distribution Strategy**, aka GTM Strategy, GTM Model, Business Model, Adoption Model. This is the most confused term. It's the chosen go-to-market model used to get products into the hands of customers. A product's go-to-market can include one or multiple go-to-market models. A company often uses multiple go-to-market models as they mature. They include:

  - *Direct sales:* a sales force is the primary source of distribution. Used most often by B2B companies with complex products and a high price.

- *Inside sales:* the customer self-serves into a sales funnel and a phone or online-based rep closes the deal. More typical for companies where customers can self-serve, have lower price points or have higher volumes of new customers.
- *Channel partners:* Leveraging independent software vendors (ISVs), value-added resellers (VARs), systems integrators (SIs), consulting firms, major regional distributors, carriers, or other technology companies for distribution. More common with highly complex products or when hardware is a part of the mix.
- *Direct to professional/customer:* customers buy products themselves, sometimes through some form of distribution (app store, physical store), and often directly online.
- *Trial or freemium:* awareness and customers come through free product usage. Customers pay for premium features if they want to access specific features or after a trial is over. In some of these models you may never be asked to pay to use the product. Happy "free" users are seen as evangelists for future paying ones.
- *Product-led growth:* customers are acquired or converted by the product itself. Often used in combination with other GTM models.

  I will refer to these as **GTM models.** When a product marketer creates a plan to bring a product to market, it leverages the go-to-market models at use in a company to distribute or encourage adoption of a product.
- **Channel Strategy**, aka Partner Strategy, Marketing Mix. See above for its use as a form of distribution. In marketing, channel strategy refers to the marketing mix across different marketing channels, such as PR, events, social, digital paid, or content. For the purpose of this book, I will specify either *channel partners* or *marketing channel mix.*

- **Product Strategy.** Connects business objectives and product vision to the work done by individual product teams. In product marketing, key elements of product strategy drive a *product go-to-market plan*, especially timing of tactics.
- **Business Goals or Objectives.** These are the specific, measurable desired achievements for a company over a set period of time. In product marketing, marketing strategies in product go-to-market align tightly with these goals.

Hopefully, this makes the relationships to concepts and language I'm using clear.

The Role of Marketing Strategies in Product Go-to-Market

Much as product marketers can't do their job without customer and market insights, no marketing activities should happen before putting in place the *why*. This is done in a product go-to-market plan, in which marketing strategies tell us the *why* behind all market-facing activities.

Declaring strategies creates guardrails that prevent activities from going off strategy. It keeps activities strongly aligned with business goals. They help teams answer which ideas are on strategy or not, reducing marketing activities that don't move the business forward.

If strategies answer the *why*, the *when* is the next most important factor. This is because whether or not an activity or tactic is relevant depends on product milestones, customer's realities, and existing market dynamics—all of which have an element of time associated with them. For example, if your target market is students, major product launch activities would be timed around Back-to-School.

The *why* and *when* in a product's go-to-market are what make the *what* and *how* worth doing. I see many companies begin their product go-to-market journey thinking of it as a list of to-dos, then asking how they should do them. Take the time to put the strategic *why* in place first.

Marketing activities also need to be grounded in the realities of a company's resources and stage. In the Pocket example, they right-sized their event to their stage—inviting just 10 customers and partners with whom integrations were deep. This exposed press to Pocket's ecosystem without taking on more than their small team could do well.

Here are some starter questions to help think through your marketing strategies. Remember, their purpose is to guide all other marketing activities. Which tactics are most appropriate depends on your answers:

- Is third-party validation important for credibility?
- What kind of customers are you trying to acquire and how fast?
- Where do those customers spend time in their professional or personal lives?
- Are you trying to educate the space?
- What are the product's strengths?
- Are there particular trends that present opportunities in your category?
- Does someone else already have established relationships with the customers you're trying to reach?
- What is the preferred way to adopt new products or technology for your customers?

A product marketer's job is to then articulate strategies specific to their product's situation and go-to-market. For example, *grow healthcare vertical adoption* or *define product suite for DevOps category*. The strategies or tactics that make it into a plan can include a wide range of marketing levers—partnerships, channels, branding, pricing, or communities—but what they're in service of is clear.

Remarkably, the strategy building blocks for a product's go-to-market inevitably fall into some variation of these themes:

- Enable growth to hit a revenue or business goal
- Improve conversion of specific customers
- Generate awareness, improve discovery, or build a brand
- Define, reshape, or lead a category, ecosystem, or platform
- Engender customer validation, loyalty, or evangelism
- Find and develop new customer segments, partners, and programs

Some worry taking the time to be strategic in a product's go-to-market slows things down or makes marketing less dynamic. Done well, it's an accelerant.

In part 3, I use a few examples to show how a strategy for one company can be a tactic for another. It all depends on stage and goals. I also introduce a lightweight product go-to-market canvas that combines all important market elements in an easy framework that makes planning across product and go-to-market teams easier.

How Company Maturity Evolves Product Go-to-Market

For a startup with one product, everything you do in market is part of a go-to-market puzzle. You're experimenting rapidly, learning about market dynamics, which customers are good ones, what product to build, and the best ways to bring it to market—all at the same time.

It's why **the product go-to-market *is* your go-to-market strategy** at the earliest stages of a company. It's also why product marketing plays such an important role at startups and why I advocate making a product marketer your first marketing hire.

At more mature companies, the go-to-market machinery is much more established and complex. The work is as much about internal coordination as it is about accelerating product adoption in the outside world. A product's go-to-market strategies might look similar to earlier stage companies but the work is done by the GTM engine of marketing and sales. Each has their own strategies and agendas, which can at times be challenging for product marketing to align. Parts 2 and 5 of the book explore much of this organizational challenge.

Regardless of where your company falls on the maturity spectrum, a product marketer is responsible for crafting the marketing strategies that shape a product go-to-market plan to keep activities aligned with what the business needs.

Once in place, you need a story that shapes how the world thinks about your product. That is where positioning and messaging comes in.

Chapter 5

Storyteller

Shape How the World Thinks About Your Product

Remember how the Word team talked about feature improvements focusing on how people actually used word processors? Or how Pocket showed how saving content to view later was part of a bigger behavior shift driven by the rise of mobile devices? Both are examples of positioning through a bigger story narrative.

Building a great product isn't enough to succeed if you don't also take the time to position it in the market. Don't make the mistake of assuming the world knows how to think about your product and why it's valuable. *You* must frame its value. If you don't do it, other market forces will.

That said, positioning a product well is much harder to do than it looks. It's more than just data, stories, claims, or a positioning statement. It's the collective outcome of everything you do to bring your product to market *over time*.

Positioning and messaging are both important and often get conflated with one another. The differences are:

- **Positioning** is the place your product holds in the minds of customers. It's how customers know what you do and how you differ from what's already out there.

- **Messaging** includes the key things you say to reinforce your positioning, making you credible so people want to learn more.

Positioning is your long game. Messaging is your short game.

Part of the confusion between the two comes from a formula that was popularized for writing positioning statements. You can find it easily if you look online for "positioning statement generators." They all churn out some variation of the following:

"For _____ who _____ the _____
          [target customer]   [statement of need or opportunity]   [product name]

          is a _____ that _____ .
                    [product category]   [statement of key benefit]

     Unlike _____ , our product _____ ."
              [primary competitive alternative]   [statement of primary differentiation]

This formula became the straight jacket of positioning. Teams took its output and applied it—as is—all over product materials. They assumed because it's called a positioning statement, they'd checked the box on positioning.

This overly simplistic approach is unhelpful to customers or worse, confusing. It's why shaping how the world thinks about a product is Fundamental 3 of product marketing and one of the most important parts of the job.

Use Formulas as Input, Not Output

Positioning starts with knowing the story you want to tell about your product—and having the evidence to support it. The most visible place this happens is in a product's messaging.

Formulas can push teams to think through their audience, a product's unique value, and the reasons to believe claims. It doesn't work well when teams rely on formulas for their final messaging. For starters, formulas tend to generate messaging that's derivative, dense, and jargon filled. It makes what something does or why to care hard to decipher.

But second, and more important, formulas focus teams on what they want to say and not on what is most important *for customers to hear.* The right level of detail or if you should orient toward technical or business value depends entirely on the audience and how well known a product is.

Let's look at two companies at the height of their head-to-head competition in the business analytics space. Both served the same audience and offered the same value proposition. One of them did okay despite having a four-year head start. The other got acquired by Google for $2.6 billion after just seven years.

Can you tell which statement belonged to which?

> "[Company A] is the best tool for running a data-driven online business. Data-driven decisions lead to better results."

> "[Company B] re-invents business intelligence. Our modern data discovery platform takes a markedly different approach to analytics. Because it operates in-database, all your data is inherently drillable and explorable."

Company A is messaging something everyone knows to be true: "Data-driven decisions lead to better results." Does that help a data analyst understand why they should pay attention to this product? It could be talking about Microsoft Excel.

While the message is simple, it does nothing to help a data analyst—someone more analytical than most—understand how better results are achieved. It doesn't say anything that might make an analyst more curious to learn more.

Company B chose to be longer and more concrete—a good choice for this audience. They are specific about what was different and say so upfront: "Our modern data discovery platform takes a markedly different approach." Then, how it "operates in-database" and what you can do better as a result: "data is ... drillable and explorable."

Even if you don't know what this means from a technical perspective, you do know what they're claiming is unique. Note how it includes elements from the positioning formula but doesn't present them in a formulaic way. Instead, they provide concrete examples of what a data analyst could do differently. It's helpful information for an analyst deciding if they want to learn more.

Company B's messaging is better *for its audience* for all these reasons. It belonged to Looker, who built a fantastic product customers loved and had good messaging to go along with it. Company A was RJMetrics; their messaging and eventual outcome was average.

Modern teams test messaging, but that's not enough to guarantee a good result. It's easy for teams to test variations on a theme and not probe the landscape of possibility enough. Testing should reveal outer bounds of what might work as well as tradeoffs in approaches.

Let me be clear: messaging isn't what makes a product good. But a good product can't succeed in the market without good messaging. It won't come from a formula; it comes from knowing what your audience needs to hear. A product marketer needs to know that to be good at their job.

A Better Process

Good messaging is honed and chiseled by multiple teams. It is never an instant masterpiece created by one team in a room. Messaging is revised numerous times with input from tests on a variety of platforms (web, in app, email, ads, and sales conversations) before it's done.

I don't advocate a formula, but offer CAST as a guide to check if your messaging is grounded in what customers want to hear. The concepts are:

1. **C**lear. Is what you do clear and is there a reason to be curious? Is being comprehensive getting in the way of clarity?
2. **A**uthentic. Is the language evocative and meaningful to your customer? Is it said in a way makes them feel known?
3. **S**imple. Is it easy to understand what's compelling or different? Will customers know what's better?
4. **T**ested. Has it been tested and iterated *in the context customers will experience it?*

Teams often iterate messaging in documents, assuming it's honed because product, sales, and marketing all got input. That is only a starting point. When messaging is tested in a web page or email, not only do you get better customer input, it's easier to see unnecessary or confusing phrasing.

Beware: simple and compelling often get confused with jargony and promotional, even among experienced product marketers. Imagine if Looker's "*Because it operates in-database, all your data is inherently drillable and explorable*" was "Because it's a collaborative data platform, there are no limits to what you can explore!" The original doesn't have trendy buzzwords, but it is much clearer for a data analyst.

I go into significant depth on this process and many more examples of great messaging in part 4.

The Tendency to Be Overly Precise

One of the biggest challenges in technical product marketing is *being technical enough at the right times versus technically precise all the time.*

It is particularly challenging for more technical products, like those in infrastructure or for developer audiences. This is where Fundamental 1 of product marketing comes into play: What does your target market most need to hear?

Messaging should be concrete enough to be credible to the technically savvy. But that doesn't mean say everything with technical precision right out of the starting gate. The job of messaging is to create a connection. This may mean messaging works best when accompanied by a product trial, videos, or customer testimonials.

Don't expect messaging to do all the heavy lifting.

Search Engine Optimization

Search engine optimization (SEO) refers to the many actions you take to improve discoverability by search engines. It applies anywhere search is used to discover something, including app stores, marketplaces, or Internet content. It is its own dedicated specialty that is constantly evolving.

Since ~70% of people's buying decisions happen online in some form, the keyword phrase ecosystem surrounding you and your competitors must be considered in how you position and message.

SEO can anchor much of your digital strategy, including content, ad buys, and even email subject lines. Keyword audits show what words others are likely to associate with your product.

Watching users search for your product in a journey test is also a quick and easy way to see how users think about your product, the language they use, and which competitors they associated with you.

Here again, SEO is all an important input, not the sole driver of the output. It guides which words to consider including in messaging. Beware the temptation to chase keyword phrases that get clicks. It can be at odds with your longer-term positioning.

The product marketer maintains a clear line of sight to the product's positioning and considers the many variables impacting it. Good judgment must always be exercised.

Positioning = Your Actions + Others'

Messaging is the most obvious artifact of positioning work, but every activity in a product go-to-market plan can reinforce positioning in some way.

If a company does a proof-of-concept during its sales process, it should direct the assessment criteria toward its strengths. For a product demo, how features are shown should reinforce the positioning. When deciding if something is worth a press release, assess if it is a good proof-point for the desired positioning. Even in very structured analyst reviews, positioning can be demonstrated with strong evidence—such as customers solving problems a particular way or repeatable tests that validate a claim.

That's the positioning work within your control.

But an equal if not more powerful force influencing a product's perception is the 70% of decision-making that happens beyond a company's control. Some aspects of this are referred to as dark funnel or dark social—engagement, buying processes, or sharing of content and points-of-view that impact adoption but aren't possible to see or track.

Comparison sites, reviews, ratings, social postings, shares, online forums, other people's content, and even employee buzz are just a search away. They collectively form a digital footprint that has enormous impact and quietly positions a product and the brand reputation of a company. Pay attention to the power of word-of-mouth and what people are saying. It can become your de facto positioning, even if your messaging and "official" marketing channels say something else.

The Long Game

I've zoomed in a lot on messaging in this chapter because it anchors how a product gets perceived. It's the starting line for a product's positioning.

But positioning happens slowly over time through all marketing actions. Persistence and consistency are key. You hold a market position much longer than any messaging. Its likewise much more challenging to change, so make sure to do this with intention from the start.

Intentional and accidental evangelists can create either lift or drag on all of this work, and they are the best way to scale go-to-market efforts. It's why enabling evangelism is Fundamental 4 of product marketing.

Chapter 6

Evangelist

Enable Others to Tell the Story

Quizlet is used every day in nearly every country of the world to study for medical school, pass beautician exams, and memorize grocery store codes. In the United States one in every two high school students have used it to do better in school.

But for its first 10 years of life, the company didn't spend a dime on marketing. Search #quizlet or #thanksquizlet on Twitter and you'll see why. Teachers around the world share photos and videos of excited students as they play Quizlet Live games. Graduates hold diplomas in hand with "Thanks Quizlet" on their mortarboards. No one asked these people to do it. They're just so genuinely grateful for what the product does, they want to share it with the world.

This is the best kind of organic evangelism that has a big impact on social media and messaging platforms. But it is just one example of an increasingly important aspect to how products get discovered and perceived—by others telling the story.

The idea is far from new. But the mediums have evolved and expanded in reach and importance. Their impact is far beyond what official company marketing channels can do.

For the purpose of this book, when I say evangelism I mean the systematic enabling of influence through others. This includes the more traditional amplifiers of go-to-market reach—sales, the press, investors, or analysts—as well as the armies of specialists in the go-to-market engine whose full-time job is to activate evangelism through their domains: social media marketers, content marketers, public relations, analyst relations, technical evangelists, community managers, field marketing managers, event managers, partner marketing, customer success, and sales managers, just to name a few.

For the product marketer, the fourth fundamental is about identifying the most important parts of this mix and applying them strategically to bring a product to market.

Enabling Others

Some evangelists matter more than others. For example, if you have a direct sales force, they're product marketing's top priority to enable. But they can't be successful if others with enormous influence, such as analysts from research firms and editors in app stores or Tier 1 publications, aren't considered equally important.

The key is understanding what type of advocacy matters most for your product's go-to-market and then figuring out what enables and then activates it.

When Word became the best reviewed word processor of its time, there was a lot of product marketing work to activate evangelism before the product was released. Some of it included:

- A detailed evaluation guide, sent whenever the product went to reviewers. A similar guide was made available to sales reps.
- Meeting the most influential pundits in person and answering their questions live.
- Doing in-depth training of the field on the sales presentation and demo.
- Partnering with product on the beta program so reference customers provided testimonials that informed direct mail messaging and ads.

Word couldn't claim leadership; pundits, reviewers, partners, and customers had to give it to us. It took consistent effort enabling the right key influencers to get there.

In the era of frequent product releases and many different GTM models, the scope of what it takes to enable evangelism has only expanded. For example, Slack's initial evangelists were users getting teammates to join Slack workspaces and immediately getting the team productive. While product-led growth is often led by growth and/or product teams, product marketing must make sure the rest of the product's evangelism efforts complement that GTM model and are effective.

Enabling evangelism for sales reps in the field means having a story and tools that let them sound like genuine advocates; they don't want to sound like they're hard selling. They also need to arm their internal champions with tools that let them evangelize on the product's behalf when seeking internal approval. I've seen many internal "sales decks" of internal advocates selling their leadership team on why expensive new software should be purchased—it's not always the rep doing that selling.

In modern customer journeys, there is a presumed bias around "official" marketing. Many search instead for what others have to say. Customer reviews, third-party blogs, social media, meetups, or the digital equivalent where more candid conversations happen is preferred. Think through if those forums help tell your story.

Communities are another common way to scale advocacy through others. It ranges from starting a group (like a Customer Council or Advisor Network), to participating in groups that already exist, to joining third-party oversight organizations. Important in any community is that it feels like genuine dialogue that's helpful for the people in it.

Effective evangelism isn't just about finding the best marketing channels; it's about reducing friction to access information that lets people effectively assess and advocate for your product. Where should a sales rep look for the most up-do-date competitive responses? Will people like or share a video more if it's on YouTube versus a company website?

As the product marketer explores the landscape of influence, understanding how to leverage it for a product's go-to-market is as important as anything else a product marketer does.

Evangelism vs. Promotion

Almost every team is good at cranking out content that talks about the product. What's often lacking is making it compelling, credible, or desirable for others to talk about.

This is most evident in product marketing that overly focuses on product. It is counterintuitive, but sometimes it's important to lead with what makes people care. Table 6.1 takes traditional product marketing tools and contrasts them with a more evangelism-focused approach so you can see the difference.

For example, let's say a product marketer is outlining a sales script for what to say in a first conversation with a prospect. The natural instinct of most product marketers is to focus on the official messaging and which features they want highlighted.

Start instead with stories on what existing customers have recently done to use the product and solve a big challenge.

Table 6.1 Comparing typical product first methods with those that make it easier for others to evangelize.

Promoting Product	Enables Evangelism
Customer case studies stating key challenges, key results*	Interest-grabbing customer stories (in video, user-generated content, photo-based on social).
Product demo that walks through features by product screens	Demo through a day-in-the-life of a customer.
Sales deck that talks about the company, product, marketecture, and key features.	Slides showing what's changed in the world that makes the need for your product urgent and important and how a customer like the one getting pitched found value.

*Case studies have a role—they just can't be the only way your customer stories get told.

This is not just because it's more believable, but also because whomever they're speaking with might learn something. Don't forget that people engage in discussions to learn, not to be lectured or sold to.

When it comes to finding the best marketing channels to enable evangelism, the modern marketing landscape is too diverse for product marketers to know all the best ones. This is a great example of where the collective intelligence of the entire go-to-market engine should be leveraged.

Tailor Evangelism Tools to Your Product's GTM

What enables evangelism depends on who is doing the talking and context. Product marketers should direct activities with that in mind. Some examples of typical product marketing-driven evangelism activities follow:

- Sales needs a well-defined playbook to engage in on-message conversations, use the right tools, and follow good process. This can make prospects fans even if they don't become customers.
- A prospective customer switching from an existing product might want to consult with a community of like-minded people so they can understand what a transition to the new product might look like. They look for a support or community forum where they can ask candid questions.
- An existing customer might like being recognized for what they do with the product. An affinity space at an event or dinner where they can talk about what they've done with other prospects can activate them as evangelists.
- Key pundits and influencers—like analysts—usually have regular cadences in which they push key reports. A product go-to-market plan must incorporate those into its timelines.

People always look for help in technology-buying decisions. In the absence of finding something about your product from a "third party" the moment they search, a competitor fills the void.

Evangelism Is a Team Sport

This work, more than any of the product marketer, is done together with product and go-to-market teams. Nearly everything is ultimately executed by others. Product marketing is the catalyst and director on behalf of the product.

Most important, product marketers use all they've learned and created in the first three fundamentals as a foundation on which this fourth gets built. But you need all four for a product's adoption.

If everything I've covered up to now feels like a big, hard, and important job, it is. That's why product marketing is a dedicated function, separate from product management and distinct from the rest of marketing.

The book's next section is where I'll dive into the specific skill set, practices, and techniques that make for a great product marketer. I'll also cover how to work with the most important partner functions: product, marketing, and sales.

Part Two

How to Do the Role: Who Should Do Product Marketing and How to Do It Well

Chapter 7

Strong Product Marketing

Skills of the Good

By the time Zack fired all but one of his sales guys, his product demo was set, he offered free proof of concepts, and he was willing to negotiate heavily on price. Yet sales weren't coming. He couldn't figure out why.

Zack was CEO of StartX (real company, names changed) which had developed new tech while at a PhD program that won a major technology competition and got endorsed by a chief information security officer (CISO) at a well-respected Fortune 50 company. Those endorsements and his team's sterling technical pedigrees got them enough VC funding to start a company.

As is common at this stage, Zack was leading almost everything—sales, product, marketing, and people—but had never done any of it before. He assumed if he wanted the product to sell, he should hire sales guys.

But it didn't work as planned. The sales guys booked meetings with people they already knew and kept asking for more leads from a nonexistent marketing team. Zack had to be at every sales pitch because his team didn't know the product well enough. He finally stopped trying to sell the product at every meeting and instead took the time to ask, "What are your top priority problems?"

It's how he learned the problem their product solved not only didn't make the top five priorities for execs; in some cases, it didn't even make the top 10.

The team realized they had misjudged the desirability and value of their product. They regrouped to pivot and solve a problem much higher on executives' priority list.

With just a few months of engineering effort, they solved a top-three problem for companies, one that intersected with a decades-old category with very mature incumbents. Just improving usability alone was considered game-changing, let alone their more effective approach.

Zack again armed his sales people with a "here is everything the new product can do" demo, but again, getting signed contracts remained hard. At this point, it was well over a year into the company's life and Zack knew it was urgent to do things differently.

He finally hired a director of product marketing. Enter Josie.

She immediately diagnosed what had been impeding the sales team: a lack of clear, repeatable messages that communicated the product's value, not just features. Along with that, the sales team needed a better understanding of which prospective customers were most likely to buy, rather than just relying on their Rolodexes. She also recognized the product needed to position itself relative to the existing category so people could connect StartX into their mental maps.

Within a few months, here's the other product marketing she did:

- A white paper articulated what wasn't working with the existing category and introduced the need for a category niche they coined. A major analyst firm was so intrigued by the white paper, they asked to be on a call to learn more.
- She created all new product collateral, bringing more consistency between the sales presentation and the website so messages were the same everywhere customers looked.

- She partnered equally well with product and sales, working in lockstep to adapt materials based on feedback.
- She worked with sales to do more in-depth win/loss analysis of the customers they were targeting and develop a more refined target customer list.
- They agreed on a product go-to-market strategy. This meant even if Josie wasn't the one doing all the work, the rest of the team knew the *why* behind everything being done by marketing.

It was a key difference-maker in StartX's turnaround. After getting product marketing in place, sales could finally do its job. They landed lighthouse customers, which made the difference in StartX getting its next round of funding.

Product discovery miss aside, Zack learned the hard way he should have brought product marketing in *before or at the same time* he brought in sales. He could have diagnosed and accelerated his product market fit issues sooner and made sales efforts much more productive. He just didn't know any of these things because he assumed to sell, you needed sales. And marketing was what you needed when you were ready to sell more.

His is a cautionary tale about why it's important to know what good product marketing can do for you, how to hire it, and how it should work with other functions. That's the focus of this part of the book.

Key Skills of Strong Product Marketers

Product marketing's purpose is to drive product adoption by shaping market perception through strategic marketing activities that meet business goals.

It requires a sharp mind and strong backbone with a skill set that overlaps a lot with product management. The difference between the roles is how skills get applied. Product managers apply them toward the creation of a product; product marketers apply them to the market adopting a product.

Here are those key skills:

Deep customer curiosity and strong, active listening. Product marketers need to grasp their customers' world. They don't arrive as customer experts; if they're not learning new things, they're not doing their job. There is no finish line for this part of the job; what happens in markets is dynamic, requiring constant processing of market and customer inputs. This is often done together with other functions, like product or sales.

Genuine product curiosity. This is crucial to earn credibility with the product team. It requires being comfortable asking questions and showing a genuine interest in the product. After a period of time, deep understanding of the product is expected. People don't need to start with product or category knowledge, but they need to have interest to learn. This same skill is what lets product marketers acquire keen competitive insights.

Strategic and strong on execution. Strong product marketers are both. But if they are more one than the other, err toward strategy as there are many others in an organization who can cover execution. That said, execution isn't just getting tasks done, it means the quality of the work achieves results. This is a form of a strategic mindset—always being able to see the bigger picture versus just getting the job done.

Collaboration. Product marketing is cross-functional by definition; it can't be done without consistent, productive collaboration with product, marketing, and sales. Nothing in the four fundamentals can be done without a lot of collaboration. Product marketers must be effective at collecting and harnessing organizational knowledge and piping it wherever it needs to go (sales and marketing). Likewise, product marketers need to consistently represent customer and market dynamics to product teams.

Strong verbal and written communication skills. These are two of the most powerful tools in the product marketer arsenal. They are used daily in every aspect of how the job gets done. If a product marketer isn't an effective communicator, they won't be clear on what a product can do for the world or work effectively with other teams. Strong product marketers are good at simplifying—they don't need to say everything to be clear. They avoid hyperbole and know how to communicate in a way that feels authentic—which is a lot harder than it sounds. Writing skills can be augmented, and presentation skills can be trained. But evidence that the person is good at communication in all its forms is key at hiring.

Broad marketing knowledge. Product marketers don't need to be experts at all things marketing, but they need to have enough knowledge of the different specialties in marketing to help guide product go-to-market activities and know why they will be effective. This requires effective collaboration with marketing partners, as they have many ideas on what can be done and are the hands that bring go-to-market activities to life. Asking questions like "How might we accomplish [insert desired outcome]?" is a great way to frame a constructive conversation.

Business savvy. This is more than just understanding business goals, it's about understanding there are different ways in which a business can grow, like new markets or new go-to-markets. It's also important to be able to recognize when an existing product go-to-market isn't working well enough. They may not be the final decision-makers, but they must understand the landscape of possibility and incorporate business thinking into a product go-to-market plan.

Technical competence. A person does not need to have an engineering degree to be technically competent; they must simply have the capacity to understand how

People and Process

technology is used. Remember, product marketing translates between those who understand the product deeply and those who do not. Being unafraid to ask questions is an essential trait to become technically competent.

It is a tall order to find someone who can do all this well, but not because the talent pool required isn't there. *Product marketers are made, not born.* What's important is setting the bar high for the role and the people you put in it.

Key Responsibilities

Product marketers are responsible and accountable for many things that require consulting others. For example, effective sales tools draw heavily from product knowledge gleaned from product managers. Likewise a customer journey map might be heavily driven by a UX team, but product marketing makes sure it reflects what happens when a customer is just starting to look.

The following box identifies some of the key responsibilities through the lens of the four fundamentals. Don't look at it as a task list, rather take it as guidance on the types of activities product marketers do to be consistently effective.

Product Marketing Foundational Responsibilities

Fundamental 1. Ambassador: Connect Customer and Market Insights
- Connect product team to market
- Segment customers and identify target personas
- Understand competitive dynamics
- Understand customer journey/customer discovery
- Understand market trends/competitive messaging and dynamics

Fundamental 2. Strategist: Direct Your Product's Go-to-Market

- Define product go-to-market plan
- Guide execution and adaptation
- Understand relevant pipeline or funnel dynamics
- Partner in branding/packaging/pricing strategies
- Guide and align programmatic marketing activities

Fundamental 3. Storyteller: Shape How the World Thinks About Your Product

- Positioning and messaging
- Frame the go-to-market story narrative/shape categories
- Create key product-centric content
- Collaborate on appropriate marketing activities and demand generation

Fundamental 4. Evangelist: Enable Others to Tell the Story

- Customers' stories
- Enable analysts, press, influencers
- Enable sales by crafting an effective sales playbook and sales tools
- Enable fandom and communities

People and Process

At the startup stage, product marketers tend to do much of this work directly, focusing on many fronts but prioritizing by what's most important for a product's go-to-market plan. At more mature companies—where go-to-markets have set grooves—this work is achieved largely in partnership with others.

Product marketers focus on both the strategic—shaping categories, growing a new market—and urgent—competitive response, sales needing training on a new release—at the same time. There is also a lot of energy spent on connective, cross-functional glue.

Beyond company stage, go-to-market models impact how product marketing gets practiced. Let me dive into some of these differences.

Growth Marketing

As a discipline, growth marketing overlaps a lot with product marketing. Both systematically understand how activities and audiences act as levers for the business. The difference is growth teams tend to be more multidisciplinary, with product resources under their direct control versus just using marketing programs to enable growth.

Growth hacking is the original term for the data-driven, test and trial, largely product-led aspect of growth. Most direct to consumer companies have dedicated growth teams, but product-led growth is increasingly important in B2B companies as well. In B2B companies, product-led growth is often described as using consumer tactics for product enabled sales.

In organizations with growth teams, product marketing will focus on bridging product teams outside of the growth team, enabling positioning, defining product go-to-market, enabling important influencers, and intersecting marketing activities with sales or marketing functions. The growth teams are more laser-focused on figuring out the mechanics, sequences, and combinations that lead to faster growth largely through the product and its data.

Direct to Customer Businesses

I use direct-to-customer (D2C) because while these go-to-markets were once the domain of businesses selling to consumers (B2C), increasingly, sophisticated business and developer products go directly to their customers (e.g. Zendesk, Atlassian, Slack, Drift). Whether the focus is consumers or professionals in businesses, this bottom-up approach focuses on a product-led, digital, or mobile-heavy approach to acquiring customers.

The product marketing work focuses on providing frameworks and guidance to marketing teams on how to appropriately engage customers throughout their life cycle, in all aspects of the funnel (awareness, acquisition, activation, revenue, retention, referral). To be effective, this requires rigor around customer, market, and product usage-based segmentation, working tightly with product teams. Thoughtful brand and pricing also matters a lot in this go-to-market model.

Focus Areas for D2C PMM

- Intensive, ongoing hyper-specific customer segmentation based on engagement and growth
- New acquisition channel experimentation
- Enabling product engagement
- Enabling customer life-cycle activities
- Understanding funnel and conversion behaviors
- Enabling important influencers

Business to Business

In B2B, product marketers discover and enable a systematic approach to convert companies into customers. Although the primary vehicle for this is direct sales, that doesn't mean product marketing focuses only on enabling them.

Modern customer journeys mean a lot of product assessment is done before a "customer" even puts up a hand showing interest. Revenue enablement means planting seeds along the buying journey. A product marketer makes sure a strategic lens is applied so the right kinds of customers get caught in the marketing net and that the right activities are prioritized especially as it relates to longer-term company strategy. For example, the primary revenue driver for a company might be an existing product, but strategically it's important that a new product get adopted. Product marketers focus on the collection of activities that enable this shift. It could be positioning, packaging, pricing, or getting advocacy in place from influencers and lighthouse customers.

Segmenting and enabling "customers" is also more complex in B2B. There is the "account" or company level, and within each account there is the buyer, the user, and many influencers. Influencers range from someone in the department using the product, to people in procurement who have no idea what the product does but are looking at what it costs, to IT who wonders how difficult a product is to support.

Focus Areas for B2B PMM

- User vs. buyer vs. influencer personas and roles in buying process
- Sales tools—competitive depositioning, demos, sales deck, sales playbook
- Understanding what nurtures/nudges people in their process
- Work with sales to define customer qualification criteria
- Guide programmatic marketing activities that align with sales targets, stages, or account-based actions

---

Product Marketing Anti-Patterns

Despite a lot of effort, there are many product marketers who aren't great at achieving the job's intent. This is largely because they aren't clear on the job's strategic purpose or aren't held to a high enough standard.

These are common anti-patterns that signal improvement is needed:

- **Strong on style, soft on substance.** Everything about the product looks professional, communicates a vision, includes benefits like "saves time and budget." But go-to-market teams keep getting questions asking what the product does and where it fits relative to other solutions. Prospects search elsewhere for "real" information or get pushed into a sales process they don't want. Sales gets frustrated that leads aren't more qualified. Product managers spend too much time on field enablement.

- **Stuck in technical weeds.** This is the inverse of the previous pattern and can be hard to recognize because what's created is technically accurate. Product materials explain *all* the features. Diagrams show what the tech does. Product managers feel like they've been able to offload a lot of the company's product information demands. But it doesn't all add up to a clear product position, leaving the competition to define what it all means.

- **Functioning like a service.** The go-to-market situation determines priorities, not internal teams. Product marketing must prioritize what sales or marketing asks for vis-à-vis what customers need. A "give them what they ask for" service orientation is seductive and common because there is swift action that makes other teams happy. But leadership is often still wondering if product go-to-market is working well enough.

- **Insufficient dedicated product marketing.** Many companies under-resource product marketing, not realizing that their market situation or product portfolio requires more dedicated focus to improve a product's adoption. This often manifests as product management doing a lot of sales enablement or teams wondering if marketing efforts are the right ones. It is often the result of a false understanding of how powerful product marketing can be for a business when done well. The function then doesn't get prioritized or people in the role don't get properly trained on how to do it well.

People and Process

Chapter 8

How to Partner with Product Management

Despite getting flamed by Bill Gates, my time as a product manager on Word for the Mac set the bar for what an exceptional partnership between product marketing and product management looks like.

Jeff Vierling was my counterpart, and we were a dynamic duo. We both stayed late when the engineering team needed help finding bugs. We went to major Apple events together to meet potential new partners and discuss their merits. When the release team—engineering, product manager, product marketer, customer support—did our regular meetings, we discussed decisions like *What are the implications if we slip the release date to fix performance issues?*

Each of us took the necessary time to understand customers and markets. One of us translated them into the product getting built; the other translated it into go-to-market. Because each of us had the other's expertise to rely on, we felt more confident in the decisions we made.

Not every relationship between product managers and product marketers is this idyllic, but it's worth striving for. Nearly every company has at least one bar-setting example of a great product marketing to product management partnership.

The product manager–product marketer pairing is *the* constant in any version of product marketing practiced. This is in contrast to the relationships with marketing and sales, which evolve quite a bit as companies mature and depend a lot on how those teams are organized.

Product marketing is the go-to-market yin to product management's product yang. Their end goal is the same—a product people love and buy—and they get there with a collaboration that lifts both functions up to their full potential.

Beyond the Core Product Team

This collaboration works best when product marketing is embedded with product teams and there is a designated product manager partner. Some refer to it as the triad expanding into a quad, but the bottom line is the product marketer feels like the product squad's designated marketing strategist.

Market strategizing and product building are different skills. It's why both jobs exist and both roles are needed. Product managers (PMs) use everything they build to achieve a product vision. Product marketers (PMMs) use everything product builds as a *portfolio of possibilities* to achieve go-to-market ends.

Translating discovery work toward what's reasonable or likely in actual go-to-market is product marketing's wheelhouse. It's some of the most important work done by the product marketing/product manager partnership.

As with the StartX founder in the last chapter, it's easy to latch on to anything that feels like it validates value, in their case, winning a contest and getting a well-respected CISO saying he loved their idea. This is especially true if feasibility, usability, and viability are already de-risked.

For product marketers, they take market learning and examine how it might impact marketing channels, distribution partners, pricing, packaging, timing, positioning, or out-marketing competitors. This is the market-fit side of product / market fit.

There is also tactical go-to-market pre-work led by product marketing. For example, if direct sales is the primary distribution channel, product marketing helps think through sales' capacity for new functionality or a new product and whether or not the incentive structure, pricing, or packaging is in place so sales can sell.

It's easy for product teams to get myopic in managing backlog, focusing on what can be built and not on how it's going to market. Product marketing is the team's partner in understanding if features or functions are marketable.

Indicators the Partnership Is Working Well

- The why behind product go-to-market is understood. PMs feel it's a thoughtful approach that matches the product being built.
- Product squads want product marketing to be involved in major product decisions to understand market implications.
- PMs and PMMs collaborate heavily on positioning and messaging. PMs are comfortable it is technically accurate and that the positioning aligns well with the product vision.
- PMs are highly engaged in analyst relations.
- Competitive response is rapid, collaborative, and coordinated.
- Pricing and packaging ownership is clear, and packaging serves customer segments and business goals well.
- PMs feel they can offload a lot of collateral and content creation as well as sales enablement with minimum consultation.

Set It Up for Success

In modern product organizations, there can be hundreds of product teams. It makes the lines of where to slot in product marketing challenging.

There is no typical ratio between product managers and product marketers. But in practice, it ranges from 1:1 to 1:5

with the average being 1:2.5 product managers for every product marketer.

Which is the right number depends on the GTM model and how much organizational support exists for go-to-market teams. For example, if dedicated sales enablement or program management teams exist, there might be fewer product marketers. If the complexity of the product is high, there might be more product marketers.

The key factors for how to align product marketing to product teams is how customers experience a product and where the business wants to grow. For example, take four products that sell into two distinct markets, enterprises and mid-market. Let's say the mid-market businesses use all four products individually and that those individual products are still growing within their respective categories. Four product marketers could each be partnered to four product manager leads for adoption and engagement for each of those products. There might be five product squads for each of the four "products" but there might only be one designated product marketer.

In contrast, the enterprise segment uses the four products very differently. They buy them as a suite, picking and choosing how they use the products and accessing them on demand. You might have designated product marketers for the enterprise. They are an overlay in addition to the individual product PMMs. They may leverage the individual product marketers as well as be designated to a product squad working on enterprise integrations. They focus on what's important for the enterprise audience across all four products.

Let's take another example that might be a huge growth opportunity for a business: building a subscription service combining many different product experiences. A product marketer would be best aligned with whichever product lead oversees how that experience is being brought to life for customers.

People and Process

Here is a brief evolution of the focus of product marketing work as products and markets mature:

Startup. This is when product and company go-to-market are synonymous. It is a time of rapid discovery and iteration, so the product marketer should feel deeply embedded not just with the product teams but with go-to-market teams. Product marketers need to rapidly adapt to what they learn with ill-defined markets, customers, and frequently adapt go-to-market plans.

Individual product forward. This is when a market and category are forged by the products defining them and credibility comes from product position and adoption. Product marketers focus on individual products. They craft the story that positions the product toward its strengths and shapes market perception. This requires a designated product marketer to product manager relationship.

Product suites forward. At later stages for multi-product companies many move to product suites. This is when product marketing shifts toward the product suite, not the individual product level, or toward customer segments, including verticals. In a larger, more mature company, it is not uncommon to have product marketing focused at the individual product level, the suite level, and customer segments. In these scenarios, there are product marketers without product manager designates. They rely heavily on other product marketers who are embedded with product teams. For product suites, product marketers are very focused on what more tightly defined customer segments need and how the suite can meet those needs. It is far less about individual products.

Verticals/New Markets. This is when particular verticals or new markets are important for business growth. Product marketers focus on verticals, and may not have designated product counterparts. Their job is to pull the story together across what's relevant from product to show how a solution is the very best and specific to a vertical's needs.

Customer segments. This is for companies where customer segments have needs that are distinct enough to be marketed to very differently. This might include different go-to-market models. Product marketers in this formation may not have designated product manager partners, again relying on product marketers who are embedded with product teams. In this model, product marketers are market needs–driven, less product driven. They are looking for market wedges that help drive adoption.

I go into more detail on how to organize, structure, and lead product marketing in part 5.

Anti-Patterns, What Better Looks Like

Because most product managers and product marketers tend to be intelligent, hard-working, and very busy, it can be hard to recognize when the partnership is not working well. Here are signs the partnership needs work and what better looks like.

- **What product ships, sales can't sell.** This can indicate product was created in too much of a "tech-first" vacuum—and there is some failure in the product team process.
  Better looks like: Product marketing is a partner in product planning and provides input on what's meaningful to the market. Beyond if a story can be wrapped around it, they should have enough market insight to challenge product teams if plans seem out of line with market realities.

- **Product ships at a terrible time to market it.** Product teams define releases by when the work is completed, not by when it can be best adopted by customers.

  Better looks like: Product marketers should always represent the market and customer perspective before decisions on the timing of major releases get made.

- **Product managers are tapped too often for product collateral or sales support.** This is evidence that the product marketer simply doesn't know the product well enough or that tools created around product aren't strong enough.

  Better looks like: There will always be a learning curve, but it is incumbent on the product manager to take the time to get the product marketer up to speed on product so this work can be reasonably offloaded. Product marketers need to know enough to be able to say what's most important and to represent it well in marketing collateral or sales tools.

PMM/PM Best Practice Touchpoints

It's important to put in place practices that work for your organization and its circumstances and resources. These best-practice processes are worth adopting in some form.

- **Ongoing: Understanding market fit.** For an early startup, this is foundational work to product/market fit. In later stages, it's a steady stream of smaller tests in market with customers and market-facing teams to ensure messaging and marketing programs achieve their desired outcomes. It's the focus of chapter 11.
- **Weekly: Product marketer part of regular product squad meeting.** Which meeting(s) depends on the product team formation and their meeting cadences, but PMs and PMMs should be in frequent contact with one another, meeting with the entire product squad an absolute minimum of weekly.

- **Bimonthly or monthly: Regular product planning reviews.** Many companies have some form of this, but it is a regular review of what's been learned in product and market discovery and what is starting to take shape in product commitments. There is more clarity on what solutions are getting created and it's easier for go-to-market teams to reflect on impact. Sales, customer success, and many other functions will be a part of these too. Product marketing should be a strong presence in these discussions and raise market implications or opportunities to take advantage of.
- **Quarterly: Review product go-to-market and product planning.** Go-to-market activities often have a longer lag-time to see effects; for example, a change in sales process takes at least a quarter to see. And even though an ad campaign has performance metrics, its impact on the pipeline might not be clear for a while. Product needs to hear what's been learned in go-to-market to ensure their planning reflects current market realities. It may change priorities. Chapter 19 goes into detail on the one-sheet PGTM canvas I recommend to drive a productive discussion.

People and Process

The importance of the relationship between product management and product marketing can't be overstated. You simply can't do any aspect of the four fundamentals of product marketing without this partnership working well. It is how a product reaches its full market potential.

Chapter 9

How to Partner with Marketing

Securities and Exchange Commission financial professionals—the people responsible for making sure all the regulatory paperwork is done correctly and on time—aren't exactly at the top of anyone's list of people you roll out the red carpet for.

But that's exactly what Workiva's marketing team did. They worked with a top-tier event production company to put together a big-budget, multiday, in-person event unlike any SEC professionals had seen. Between sessions in which attendees could earn continuing education credits, they could Spin-to-Win and walk away with a guaranteed prize, make-a-sundae where they got to keep the mason jar, or speak with an army of eager product managers and product marketers answering questions and listening to suggestions. Each evening featured experiences like a social in an aquarium or exclusive access to a theme park.

Attendees left the conference *raving* fans, engendering loyal word-of-mouth in a community not really known for that. It even swayed skeptical analysts who couldn't ignore a company with such passionate fans. The conference paid for itself every year in converting customers.

This annual event helped Workiva punch above its weight in the market. It's the kind of coordinated, branded, big-impact event that's a great example of what marketing teams can do so well. Their collective work creates the experience customers associate with a company, and that paves the way for successful go-to-market.

Using the Right Marketing Mix

A quick reminder that while product marketing is often a marketing function, in this chapter, when I say marketing, I'm referring to all the other roles in marketing.

Marketing brings much of go-to-market to life, which means they play an enormous role in shaping a customer's experience of a company—from things as small as an email signature to major events like Workiva's. In the marketing world, the superset of how a customer experiences all these touchpoints is referred to as brand experience (more on this in chapter 16).

Not everything a company does in marketing is about a *product's* go-to-market. Marketing specialties like brand, public relations, events, social, or demand generation live at the company level, serving all products and the company's overall goals.

Product marketing makes sure the right work is happening in this machinery for their product's go-to-market. They define the strategic palette from which a product's go-to-market activities get drawn, then guide their execution. The marketing organization relies on the product marketer to make sure they get the product part right.

The messaging foundation led by product marketing— often tested by marketing—is how the marketing team knows what to say about a product and in which circumstances to use it.

The strategy foundation led by product marketing in a product go-to-market plan is how marketing teams know the *why* and *when*. The teams collaborate on the *what* and *how* that best match a product's go-to-market goals.

People and Process

It's not always obvious to those outside of marketing how a particular activity connects with a product go-to-market. This isn't a comprehensive list, but it connects what a marketing team is trying to do with some of the typical activities done (for a detailed explanation of any of these marketing terms, please see the appendix).

- **Making customers aware of the problem, the solution, and a company.** Frequently used activities include advertising—traditional (TV, radio, print, outdoor) or digital (mobile, search engine, or display)—website, search engine optimization (SEO), press articles, or analyst reports.
- **Encouraging consideration of a solution.** Frequently used activities include white papers, videos, customer stories, events, email, account-based marketing, direct-mail, partnerships, public relations, and press articles.
- **Encouraging purchase or renewal.** Frequently used activities: pricing, packaging, customer events, reviews. Not marketing but part of the mix is customer success.
- **Enhancing awareness and loyalty around the brand.** Frequently used activities include customer community, social media, content, newsletters, influencers, third-party events, and press articles.

A broad, well-coordinated mix—not any one thing—is the winning formula for good product go-to-markets. Often, it takes deliberate experimentation to find the mix that works best. And that mix changes over time as products move through the adoption curve.

That's why the ongoing collaboration between product marketing and marketing matters so much.

Indicators the Partnership Is Working Well

- Marketing understands the nuances of the market and how to best segment customers. Marketers feel they know key insights that help them do great work.
- Marketing understands the context for why a product is valued, not just what features matter, and has a strong messaging framework that lets them do their jobs well.
- Marketing understands the why behind recommended activities, and a variety of new ideas are explored to reach new or existing markets. Marketing teams do more than just double down on what has worked in the past.
- Decisions around product naming shifts or moves toward a line of business brand are done collaboratively and with the larger company brand in mind.
- The cost to acquire customers is sustainable for the business.
- Teams collaborate on how to adapt product-related marketing in response to shifts in markets.
- Product marketers feel their product's go-to-market goals are being well served by actions marketing takes.

People and Process

Set It Up for Success

Product marketing is the product's ambassador to the marketing team. They ensure marketing stays aligned with product strategy. They work with marketing to broaden the range of activities used for a product's go-to-market.

Bringing variety to the marketing mix and trying new ideas is an important part of the marketing–product marketing partnership. The product marketer is constantly trying to find new ways to expand product go-to-market, while the marketer is constantly trying to use product to improve marketing outcomes. By working well together, everybody learns faster.

In most organizations, product marketing reports into marketing, which creates a natural presumed alignment between teams. Some organizations have product marketing reporting into product (I'll explore this and its pros or cons in more depth in chapter 26). But shared reporting structure does not guarantee good collaboration. It's important to set up processes that let product marketing be systematic in how it coordinates with marketing.

Agile marketing, an increasingly popular practice in marketing, is one of those methods. Led by a product marketer, they lead weekly meetings with marketing specialists to discuss and agree on how to prioritize the workload and what should be learned from recent campaigns and activities. I'll go into more detail about this in chapter 12.

Anti-Patterns, What Better Looks Like

Although marketing teams execute much of a product's go-to-market, they often don't know when they're missing the customer or product context inspiring it. Execution may get out of step with what customers truly do or how they feel. This is where marketing teams must balance campaign clicks with *does this send the right message?*

- **Campaigns perform but product isn't getting positioned.** Every campaign has performance metrics, and it's easy to get caught up in maximizing their individual performance. A lot of money can be spent in a short amount of time—particularly with digital—that doesn't necessarily translate to what the business needs. Whether or not performance is good is always relative to intended purpose. *Better looks like:* Product marketing is part of the planning process for campaigns to make sure—in aggregate— they help position the product appropriately, say the right messages, and stay mapped to the company's goals.

- **"Future state" is too far ahead of current reality.** I've seen marketing try to solve slow velocity or lack of differentiation by pushing hard on a future aspirational state that is so far ahead of product realities, it detracts from a company's credibility. Finding the right balance between inspiration and believability is a hard but important line. *Better looks like:* Product marketing is the ideal partner in finding the line between aspirational future that inspires versus where current product reality should play a bigger role.
- **Creativity that pushes the edge but doesn't connect with customers.** Sometimes marketing falls in love with an idea that doesn't feel true to the target audience but is great at getting attention. Clever always has a place, but the most clever also connect to the product too. *Better looks like:* Product marketers are the eyes of their audience. Before making any decisions on if a bold new idea will fly or flop, leverage testing heavily by putting ideas in front of customers to get their reaction.

PMM/Marketing Best Practice Touchpoints

As with product, do what works for your organization and its particular situation, but consider these best practices.

- **Weekly: Alignment meetings.** This is where marketing and product marketing come together to ensure the right campaigns or assets are being prioritized. There can be shifts in product, or something more urgent needs to jump to the head of the line, for example, a swift, coordinated competitive response. It's also a time to look at campaigns to see if fine tuning is required. I'll talk about how to do this using agile marketing practices in chapter 12.
- **Monthly: Activity reviews.** Monthly is a good cadence to look at what's been learned that should be applied into what's planned for the subsequent month. It's also a good cadence for reviewing funnel metrics, which often take time to

(continued)

reflect any changes. This is how product marketers and marketing know if what they're doing is adding up to the intended business outcome.

- **Quarterly: Revisit product go-to-market and funnel metrics.** Marketing teams need updates to any shifts in product priorities. It's also a key time to tie what's happening with sales results into where to focus next quarter's work. The one-sheet PGTM canvas in chapter 19 is a great anchor for this work.

The marketing–product marketing partnership is an energizing and crucial partnership in how go-to-market scales. Great partnerships help marketers do their best work and enable businesses to reach their full potential.

People and Process

Chapter 10

How to Partner with Sales

When this company's midmarket sales team badly missed their sales target, they arrived at their quarterly review meeting with only one request: paper data sheets.

This was *not* 20 years ago—it was just a handful—when digital only product collateral was the norm. The marketing team was stunned. Not only was no other sales team asking for it, but it was this sales team's only ask.

When pressed, the team said they needed a leave-behind that prospective customers brought back to their desks and that kept the product more top-of-mind than an email. Plus no one else was doing paper data sheets so it stood out.

This sales team was notorious for going after companies that weren't quite the target customers the sales playbook was built for. The marketing team asked if the wrong customers were being targeted and if that was a possible reason for the team's shortfall. The sales lead rebuffed the question, "No, we just need paper data sheets."

The problem was not that one team was right and the other wrong. It was that there was a demand, not a discussion, in which sales presumed the solution (paper data sheet) instead of asking for what they needed: a more memorable prompt encouraging follow-up. It also didn't bode well for the marketing-sales relationship that marketing couldn't ask about how prospects were being targeted without sales getting defensive.

Much like a product manager takes customer input to create the best product for what a customer is trying to do, product marketing does the same for sales and marketing. They take sales input and work with marketing to create the best response for what sales is trying to do. They likewise represent sales needs back to the product team to help prioritize product work.

In this particular case, marketing had many other ideas that weren't paper data sheets and met sales needs: a customized video email, a direct-mail follow-up, or additional training on ideal customer profiles. All of these should have been discussed.

Highly coordinated marketing and sales activity is now table stakes for successful go-to-market when a salesforce is involved. Anything less doesn't break through the noise. Put simply, sales can't meet their goals at the speed they need without a great partnership with marketing. And that starts with a great partnership with product marketing.

Balancing Urgent and Important

Sales wants to know what to say to whom to drive a deal forward. Their incentives are around closing deals so they make their numbers each quarter. For that, they need pipeline and training. Marketing is the means to that end.

Product marketing is the gate to the necessary product knowledge key for selling. Specifically, they shape it for the market, versus something totally raw coming straight from product teams. Product marketing also creates the deeper product content and tools that help sales sell. They then direct across marketing specialties to make sure marketing activities add up to a product go-to-market that lets sales sell.

A natural tension exists between sales and product marketing. Sales wants to get it done now. Product marketing wants to get it done right.

The most powerful tools in this relationship are (1) reference customers and (2) a sales playbook. The first is necessary for the second because it provides the concrete, real-life examples from which best practices for success are discovered. Once clear, those practices are written into a playbook so they can be repeated by others. Product marketing then ensures sales is trained on what's most important in the playbook: a demo, competitive response, or how to run a proof-of-concept process.

Playbooks show appropriate actions by stage, next step, relevant tools, and what criteria qualifies a customer to flow from one stage to the next. Product marketing drives its creation at earlier stage companies, but it is always developed hand-in-glove with sales as they confirm what repeatable success looks like. It connects relevant tools created by product marketing to the sales process. When done well—and if followed—a good playbook can make an average sales rep successful in a faster than average amount of time.

Managing how to prioritize other marketing activities comes from looking carefully at funnel metrics (like conversion rates and time in stage). Product marketing works with marketing to examine each stage of the customer funnel and see where adjustments are needed.

In general, when a direct salesforce is involved, prioritize adjustments from the bottom of the customer funnel before moving up to the top stages. If customers aren't converting at the end of the process, there is no point in more prospective customers starting it.

Key activities product marketers work on with marketing in support of sales include:

- Jointly defined ideal customer profiles, target customer segments, or target account lists
- Customer journey maps
- Creating sales tools, including sales presentations, call scripts, email templates for prospecting
- Product data sheets or videos, product information on the website

People and Process

- Product demo(s) grounded in product's key messages
- Competitive response tools
- Ensuring customer stories or case studies are usable
- Customer advisory board
- Identifying new target markets
- Identifying key events important for the product's position and target customers
- Sales training and enablement—and working with sales on how to stay on target in who it sells to

While product marketing sets the strategic framing for this work, the marketing and sales teams do a lot of the execution.

Indicators the Partnership Is Working Well

- Sales is knowledgeable about the product and targets the right ideal customer segments.
- The sales playbook is being followed.
- Healthy flow of prospects through the pipeline, and whenever gaps are identified, work with marketing to select activities that close those gaps.
- Materials connect well with intended audiences and articulate the pain people feel today. Product marketing helps marketing stay grounded in what creates urgency while being appropriately aspirational.
- Marketing actions and assets are relevant, timely, and compelling. These can be content pieces, press, or an analyst report. Product marketing ensures sales has a wide array of marketing tools to choose from to help push their deals forward.

Set It Up for Success

Sales is the one to one, human part of go-to-market; marketing the one to many, scalable part. Any good sales–marketing partnership doesn't solely rely on individual leadership dynamics to succeed.

Product marketing needs systematic ways to collaborate regularly with sales. This starts with how leaders empower teams to collaborate as partners. It also requires team interfaces to be defined. For example, product marketing should attend weekly pipeline reviews to hear what's working and what's not. This helps them adjust marketing response to sales realities. As companies mature and sales becomes more predictable, this cadence or forum evolves.

But like the StartX example showed, sales can completely fail without good product marketing. Clear customer segments where there is higher likelihood of success must be identified. Reference customers that prove the product works as claimed are necessary.

The best tools to make sales successful are created using a collaborative process, with input and testing from sales. Sales should feel armed and prepared to respond to current market situations they experience daily.

Sales also needs pricing and packaging that makes products easy to sell. Packaging is most often led by product marketing, which also has a heavy hand in pricing (see chapter 17). Both need to be easy for sales to explain with a mental model that is easy for customers to understand. Sales often handles negotiations with procurement departments and needs to know how to speak to value, not just price.

Anti-Patterns, What Better Looks Like

If product is the engine of growth, sales is the gas. Marketing is the gas station. Product marketing is the gas station attendant making sure sales gets tanked up with enough to get to their next stop.

When this partnership isn't working well, sales lag expectations. The root cause can be a mix of marketing and sales, but bad patterns to avoid include:

People and Process

- **Marketing as a service to sales.** If marketing serves sales requests without product marketing guidance, it becomes easy for activities to only focus on immediate and not also strategic needs.

  Better looks like: Examine what moves toward product go-to-market goals and use data to drive prioritization.

- **Lack of awareness before sales engages.** Especially in an era where the vast majority of customers' journeys are self-directed, if a product, problem, or company isn't anywhere on someone's radar, it makes sales' job really hard. "Spray and pray" awareness style campaigns are extremely costly and only appropriate for more mature organizations that have the scale and need for such broad awareness.

  Better looks like: More targeted approaches in which marketing focuses specifically on customers or accounts where sales is engaging. Account-based, customer-journey coordinated approaches are great here.

- **Lack of adherence to sales playbook or use of official materials.** Sales taking maverick actions to get deals done is detrimental to the company. Customers that aren't good matches for the product can't be retained. Inconsistency in messages damages the product's ability to hold a position. It hurts the entire go-to-market machinery of a company.

  Better looks like: If tools are routinely not being used, it's a sign they don't meet the sales teams' need or that sales isn't being held accountable to position products in a consistent way. Playbooks and the tools used in them work best when created collaboratively. Ideal is to select a group of sales reps to work with product marketing on sales tools for a set period of time to make sure it all works before getting broadly rolled out. Likewise, adjustments in strategy, messaging, and activities occur as a result of this test run. Sales management should then hold reps accountable to use playbooks and tools.

PMM/Sales Best Practice Touchpoints

As always, adjust this to your organization and its particular situation.

- **Weekly or Biweekly: Pipeline and marketing reviews.** This can be participating in weekly sales meetings/calls where pipeline and prospect activity are discussed. But depending on the size and configuration of your organization, it can be its own meeting. It's a place for direct dialogue about what marketing activities can get better aligned with sales efforts. Product marketing helps determine the best ways to respond strategically: Is there a product deficit? Is it a positioning or messaging adjustment? Is there an information or training gap with sales?

- **Monthly: Joint funnel analysis.** Ideally already in place with marketing, this is a place to diagnose if targeted work—a new competitive asset, an e-book on a timely topic—can improve effectiveness for either sales or marketing. Monthly examination lets teams see what is actually having an impact. Likewise, marketing often establishes clear service level agreements between marketing and sales on what constitutes a handoff between the two functions. Product marketing watches performance around these handoffs to see what adjustments might be needed on either side.

- **Quarterly: Revisit product go-to-market and funnel metrics.** It's important for sales and marketing to have candid discussions based on data about how much growth in the pipeline is needed for the business and how much will be driven by marketing. Product marketing ensures this informs how product prioritizes decisions. There may also be additional marketing work in support of sales that is not directly guided by the product go-to-market strategy.

People and Process

Any company with a direct sales team can't succeed without the vital partnership of product marketing. In the best partnerships, they provide the foundation from which salespeople are able to do their best work.

Chapter 11

Discovering and Rediscovering Market Fit

There is one profession required—by law—to do their jobs using the written word: lawyers. Formatting as specific as how to indent a subpoint in a legal brief is strictly defined. And back when I was on the Word team, it didn't do it right.

My last year on the team, not only was it *not* the de facto standard for the legal market, it was the one remaining foothold of our erstwhile rival, WordPerfect.

Unsure of why, the product team did a national tour of law firms not using Word. They examined hundreds of documents and did dozens of in-depth interviews. They learned there were formatting needs for specific legal briefs that Word couldn't do without major workarounds.

Fixing these issues required a rewrite of some of Word's underlying layout engine. It meant the "legal will love this" version was at least a major release cycle away. Even though it was our biggest growth market, we couldn't fully hit the gas until those product issues were addressed.

The product marketer dedicated to legal came up with how to fit the current version to market realities: focus marketing on law firms who were more technology forward. For example, those who wanted to use the rest of Microsoft Office—PowerPoint for trial presentations or Excel to make charts.

Word has now long been the standard in the legal market. But it shows that even for mature products, product/market fit is not a one-time thing. As products and markets evolve, the product needs to respond to the customers it wants to grow into. The go-to-market approach must respond to customer and market realities as they are. Each shapes the other.

It's why discovering and rediscovering product/market fit is one of the most challenging parts of bringing a product to market *and* the most important to do well. Doing the work is not the exclusive domain of product marketing, but understanding and applying it to a product's go-to-market is.

The Market Side of Product/Market Fit

I see so many companies find an initial beachhead of customers, think they have product/market fit, and then find their growth stalling. It's never about just one thing. But it's often because not enough attention was paid to market fit. As Word in the legal market shows, discovering the product and its market aren't independent activities. They are discovered in parallel.

When I say market fit, I mean discovering "market pull." What makes a customer need or want your product enough that they *take action* to learn more, try, or buy. *And* what makes this pattern repeatable.

Beyond a product's usability, feasibility, and viability, product discovery work around value is intended to find this answer. Although it is the most difficult and important risk to probe, it tends to be underdeveloped.

People *say* they will use or buy a lot of things, which is easy to confuse with product desirability. Not all tests and techniques are equal in placing decisions or actions into a realistic enough market context.

Determining market fit is about taking inputs from discovery work and applying them to what people are likely *to truly do* in real-life conditions. How will they act when confronted with

People and Process

a crowded market, competing priorities, limited budgets, and a status quo that works well enough?

Value discovery work must probe beyond "would you buy this, would you buy it from us, and how much would you expect to pay for it?" Assessing market fit requires probing more deeply on market conditions that motivate action or create urgency. Sometimes it can be about how a product is distributed.

I see many companies struggle with how to take what they're learning and understand the market implications. It is why a product marketing partner is so important in this work. Good product marketers can apply more market nuance to doing and applying discovery work.

They catch things like segmenting customers by those that might be better evangelists and not just focusing on likely initial users. Or how actions might be more influenced by word-of-mouth from what is read on a comparison site or on a developer forum than an actual trial product experience.

Synthesizing all this learning into a smart product go-to-market plan, positioning, and messaging is the work of product marketing. But it's important that strong discovery work happens first to inform all of it.

Probe Early, Probe Often

Traditional market research from analysts, third parties, and trend reports is always good context, but nothing beats live customer conversations (calling them interviews can be a little intimidating) for understanding market context.

Discovery work led by product managers should always cover product baselines:

- Are your customers who you think they are?
- Do they really have the problems you think they have?
- How does the customer solve this problem today?
- What would be required for them to switch?

The market fit side of the equation explores what lies underneath perceived value. It probes market dynamics affecting what people think, what drives growth, and what creates urgency.

The questions below don't need to be asked all the time—nor are they the only ones that might be appropriate to ask—but the collective discovery work done should answer these more market-oriented questions.

Value
- Who is most likely to use this? Who buys it? Who influences decisions?
- Do they prioritize the problem?
- What else are they considering that is similar?
- What have I said/shown you that is most compelling (aha moment)?
- If probing relative urgency: If you had 10 points in your budget to allocate to solve your most urgent problems, how many points would each get?
- If probing pricing: How much would you expect to pay? Maximum price willing to pay? What else have you recently bought at this price?
- What new product have you most recently bought and why?

Growth / Connection
- What created curiosity?
- How would they describe this to a colleague? (crucial for messaging insight)
- Where would they expect to see this product talked about?
- How would they expect to assess it?
- What would make them a raving fan?
- Who are the next most likely market segments to adopt? Do they feel the same sense of urgency? What market situations are necessary for them to act?

People and Process

Any product discovery technique that engages directly with customers can be adapted to inform this work—from prototypes to usability tests to A/B tests to the Sean Ellis test. I also advocate using simple market test techniques that give quick, directional feedback and illuminate how people think, feel, and, most important, act.

Here are some examples of discovery techniques and how to use them to probe market-fit dynamics:

- **Exit surveys.** For users who immediately exit a website, use any number of tools to pop-up a one question survey: "Is there anything we could have done differently to make you stay?"

- **A/B messaging test.** There are many tools that make it easier to test and optimize product descriptions or text on websites. When you do A/B testing of messaging, don't just pay attention to what performs better. Ask what the relative performance differences tell you about market perception. Are you positioning yourself in the best category or to the right customers to be valued?

- **Demand test variation.** On a website where a click leads to a landing page demand test, ask *only* for some sort of buy or trial commitment. This separates the curious from those actually willing to take meaningful action. You must be transparent on the state of your product (if it's an idea versus an existing product) but you can still ask "Why did you decide to buy?" right on the page in addition to asking if people are interested in follow-up.

- **Ad testing.** Great for messaging or seeing engagement on social or search platforms. I recommend testing vastly different directions, for example, aspirational-, product-, or problem-oriented messages. You're not trying to optimize which is best; rather, you're trying to learn what direction gets people to act and what their relative performance tells you about the market.

- **Sentiment probe.** Take a baseline sentiment measure (use a 7- or 10-point scale) on interest for your product, problem, or space. Then show a video—your product, a competitive product, an ad, a brief explainer video—and measure sentiment with the same question after the video to see if there is any shift in interest. If there is, ask those who shifted why.

- **Usability test variation.** In addition to what you normally probe, include key competitors' websites to watch users navigate through and ask what actions they would take next. Another variation is watching people search for your product or one like it in the space. This helps you understand a more complete customer journey as opposed to just your product's in isolation. It's better at shedding light on what people are likely to do as they explore a problem and the space in which your product operates.

I encourage people to be creative and get out of the building to really understand market fit. That's where a lot of big "aha" insights happen that might otherwise go unnoticed.

Creative Market Test Ideas

Running quick market tests is the final project of the course I teach at UC Berkeley on Marketing and Product Management. Students have to run at least three in a week and assess their implications. Here are some of my favorites:

Product: a tool for developer application security

> **Market test: In-person messaging.** Putting up signs with different messages at a mass transit train stop and tracking passerby glances and walk-ups as indicators of interest. The one with the most glances didn't get any walk-ups. The one with fewer glances did get a couple walk-ups who asked questions.
>
> *Market implication: The more specific message was less interesting to most but more interesting to the right target audience.*

(continued)

Product: wearable wristband sleep tracker and coach

> **Market test: Short in-person survey in front of Walgreens at rush hour.** They then came back the next day at the same time but with an actor in a doctor's lab coat—who remained silent—nodding behind the surveyor to see how much the implied endorsement affected interest and likelihood to try the product.
>
> ***Market implication:*** *A perceived medical endorsement noticeably affects people's perception of the product's value and interest in learning more.*

Product: online bootcamps to get people job-ready for careers in technology

> **Market test: Value, pricing, and brand perception test.** Value matching game where participants were asked to select from five different dollar amounts and assign each to a different company that was competing in the same space. They then compared these answers to how participants had answered questions about how important a well-known brand was in making decisions.
>
> ***Market implication:*** *People like to think they aren't affected by brand reputation. But when forced to make choices compared to similar products, they are willing to pay more for the brand they know, even relative to what they feel that brand is worth in a different context.*

Additional Techniques to Understand Market Fit for Existing Products

As Word in the legal market showed, market fit is a function of product and market realities together evolving over time. When you have an existing product in market, it can be particularly hard to recognize when you no longer have market fit. These additional techniques can reveal if market fit might no longer be there.

- **Win/loss analysis.** It is crucial to understand why you win and lose customers. It's not just who you're losing to (although that matters), it's the why. Is it product or process? How much is perception or brand part of the mix? Win/loss analysis gives you insight on what might need to be reframed or messaged better, where product may need to improve, or where sales process may need to change. Product marketing's unique relationship across all these functions is what makes this form of discovery work so important for them to drive.

- **Sales call shadowing.** Whether listening to a recording of sales calls (many sales intelligence platforms do this) to being on one in person, there is nothing like seeing or hearing the direct interplay between sales and a prospective customer. There is body language and intonation that can't be communicated in text and speaks volumes on sounding 'pretty good' versus a clear 'wow' but isn't said directly. These important nuances are critical to understand, especially in messaging but also in making sure the right customer segments are targeted.

- **Intent data.** Many tech companies have go-to-market data teams that have in-depth access to customer data from account-based or predictive marketing tools. These platforms can reveal what marketing actions prospective customers take that historically translate into sales. Some products combine those insights with technographic, firmographic, and account actions. Product marketers can leverage these insights to help with market segmentation, messaging, and assisting marketing teams with campaign development to make sure a product's go-to-market stays in line with how customers behave in the market.

- **Social/customer sentiment.** Sentiment can't be ignored. From customer product reviews and postings on social media to how employees talk on sites like Glassdoor, all of this adds up to how people really feel about your product and company. Both positive and negative sentiment inform what you might need to work on.

People and Process

Reputation might lie in how your company handles customer support. Take these social signals for what they are—directional feedback that shows how people feel, not necessarily how things are. But know that belief equals truth for many people.

Timeboxing

Perfection can quickly become the enemy of great. "We don't yet know who our ideal customers are," or "We don't yet know which messages are most compelling." A period of discovery and trial is essential, but it is equally essential to not wait for definitive data to make the right choice for right now.

Timeboxing is a very simple but tried and true technique of allocating a fixed time period for a planned activity, then assessing whether or not you've reached a reasonable outcome at the end of it. It establishes a time boundary at the start of the process, not as a result of the process, and keeps you accountable to moving at the speed markets require.

Whenever doing customer discovery work, I encourage timeboxing the learning periods so there is a time for rapid learning, then some time spent executing and iterating on the learning. Anywhere from a week to a month can be enough to create a starting point.

Active Listening

There is one particular quality that distinguishes those who are great at product and market discovery work: active listening. They don't listen to respond or have their assumptions validated. There is open, attentive listening; their mission is to learn.

Lead with open-ended questions that reveal market perceptions before diving into specific product discovery ideas. If you lead with your ideas, you've primed the response. It's easy for people to now be anchored around how you're thinking of the solution, not how they would.

Equally important is to put enough different ideas in front of customers so you can gauge the relative strength of response. I encourage always using some form of calibration, like ratings, rankings, or even "Would you hit the like button?"

If you're a product manager or a product marketer thinking "we need to do more to probe market fit," you're right. Both roles don't tend to do enough and need the insights to do their jobs well.

Next we'll focus on the opposite end of the spectrum—getting finished products out to market—and how to keep everyone aligned in the age of agile.

People and Process

Chapter 12

Product Marketing in the Age of Agile

When Jade became the team's product marketer, she was excited to prove herself. She listened intently at her first product team standup and was quick to point out a feature that would be good to actively market.

At the standup the following week, she expected to hear an update on that feature. When she didn't, she asked the product manager and was told they had already released it, as documented in last week's release notes.

Shocked, she asked why anyone hadn't told her. The product manager replied they had—in the release notes. Jade scrambled the marketing team together to email existing customers and do the social promotion she had planned.

On the flip side, Jim, an engineering lead, had several teams hard at work over multiple release cycles to do large performance and stability improvements. The engineers waited to see something in the company's marketing about the product's now blazing performance but saw nothing. Jim wondered, after all this engineering effort, what gives?

This is the challenge of product and go-to-market alignment in the age of agile. The velocity, lack of predictability, and lightweight forms of communication and documentation make it hard for both sides. It is particularly challenging with continuous delivery of product. Go-to-market teams rely on more predictable cadences and planning to do their jobs well. For product teams, what gets marketed and why doesn't always make sense or correlate to effort.

The fix lies in clarifying expectations and defining a process that lets all teams categorize "releases" together and act accordingly.

Agile also brings an opportunity to rethink how marketing gets done. Agile marketing practices are growing in adoption. They take the same agile principles used in product and apply them to marketing. It enables a more dynamic approach that responds to market shifts quickly. In both instances, the product marketer is key in making the models work.

This chapter tackles the tools and techniques for both.

Create a Release Scale

First, some terms.

A sprint is a short, time-boxed period when a product team works to complete a set amount of work. It may or may not mean providing access to that new functionality.

A release makes public a new product or combination of features that provides value to customers. It usually includes some coordinated go-to-market action across the company. Sometimes people call releases launches. This is because the term is used loosely in real life.

For go-to-market teams, a launch is a major, cross-functionally supported release that the entire company gets behind. It's usually oriented around a set date that allows support from every interaction point associated with a product's release. It creates a focal point for coordinated activity so efforts have outsized go-to-market impact. Most companies won't attempt more than one to two major launches a year.

The distinctions between a minor release and a major launch are really important for everyone on the go-to-market side. What does or doesn't get done flows from how releases are categorized.

Developing a shared understanding of how a release should be categorized and what go-to-market activities will get done is the purpose of a Release Scale. It's a simple tool whose sole purpose is to create shared vocabulary and expectations between the product and go-to-market teams (Figure 12.1).

This is different from a detailed organizational plan for every action needed for a product launch.

The Release Scale sets standards around release "types" and the go-to-market activities done to support them. It clarifies the broader marketing objectives of any particular release or likewise if one requires more marketing attention because it has high customer impact. It also clarifies how much lead time marketing teams need to do their job well.

Product marketing is the focal point to make sure these conversations around releases happen with product. They also often manage across go-to-market functions to ensure everything necessary for a release is done in a coordinated way.

The teams that make the work listed on a release scale come to life include marketing, customer support and sales enablement.

As companies grow, many have broader go-to-market or regular product planning meetings where the categorization of releases are discussed by all affected parties: sales, customer support, marketing, product, engineering, operations.

When something rises to the level of a release, there is open debate about what category/level release it is. That then determines the amount of resource and support it commands from the go-to-market side.

	Level	Example Releases	Go-to-Market Goal	Typical Resources	Lead Time	Cycles
Low	1	• Customer fixes • Minor performance fix • Minor mobile release	• Customer satisfaction	+ Release notes + Blog + Twitter	Continuous	Weekly
Medium	2	• "Sexy" feature	• Category validation • Customer delight	*All listed above plus...* + Feature benefits on website/campaigns + "New" social campaign	1 – 2 weeks	Monthly
Medium	3	• Industry trend • Partner integration • Internationalization	• Important to select markets • Industry validation • Important to win deals	*All listed above plus...* + Select audience campaigns + Partner engagement + Competitive campaign	4 weeks	Opportunistic
High	4	• Partner announcement • Major industry (iOS 8)	• Broad awareness • Industry validation • Category leadership • Customer validation • Improve sales or up-sell	*All listed above plus...* + Fact sheet + Short lead PR + Customer campaigns + Coordinated partner promo + SEO/App store focus + Sales training + Event support	6 – 8 weeks	Quarterly
High	5	• Major launch • Rebrand • New product line	• Lead industry • Mainstream awareness • Grow revenue streams	*All listed above plus...* + Paid ad campaigns + Advanced PR + Dedicated event	3 – 5 months	1 – 2 / year

Customer Impact

Figure 12.1 Release Scale example.

Here's how to create your version:

1. **Decide on the calibration scale.** Be it levels, grades, names, numbers, or tiers, I encourage using something that doesn't require a legend for people to understand. Stick with hierarchies that have a clear mental map for the differences in levels.

2. **Use known past releases as examples to define the levels.** This is an important and often missed step. Don't make this an academic exercise of a potential future release. By using known reference points, people have a known comparison to calibrate from.

3. **Identify customer impact.** Most changes don't impact customers in potentially harmful ways, but some do. The degree might shape the level on a scale. Keep this simple: low, medium, high.

4. **Define marketing objectives.** What is the market purpose of this release? Is it catching up to the competition? This again helps people place a release appropriately within the scale and makes clear why a small release might merit a Level 2 whereas a similarly small release is only a Level 1.

5. **Define typical resources and promotional vehicles used.** This is important in getting product teams more knowledgeable on the range of promotional levers used by marketing. When is something worthy of a paid campaign versus an email to customers? When does a video get created? In this example, Level 1 defines the minimum amount of go-to-market work a team will do. Each subsequent level lists what gets *added* to the levels preceding it to bring it to market.

6. **Articulate needed lead times.** People don't understand lead times required for marketing activities unless they've done them. For example, website updates, legal

reviews, copy editing, design, pitching press, getting customer support ready can all take weeks. Make visible the lead time necessary to do the work.

7. **Talk about releases using the scale in planning meetings.** Any time there is talk of a key release, bring in the scale. "From the product team's perspective, is this a Level 2 or Level 3 release?" Also, remind them that it's important for the company to have one or two Level 5 releases a year. What can be done to give the company an important go-to-market moment? Companies that lack big visible movements forward start to give the perception they've stagnated.

Without exception, every company I've encountered that uses some type of release scale says it vastly improves communication and expectation management between the product and go-to-market teams. It's an incredibly easy tool that leads to meaningful improvements in go-to-market execution—and less needless frustration.

Agile Marketing

In recent years, I've seen a rise in the practice of agile marketing. It adopts aspects of agile—continuously prioritizing and adjusting—in the approach to how marketing gets done. The basic principles are:

- Respond to changes versus following a plan
- Rapid iterations over Big-Bang campaigns
- Testing and data over opinions and conventions
- Numerous small experiments over a few large bets
- Individuals and interactions over "large" market segments
- Collaboration over silos and hierarchy

The highest functioning version of this has product marketers act like product managers with a marketing squad. They lead this cross-functional group of specialists—communications, advertising, digital, web, design—in weekly "scrum" meetings to discuss how to prioritize workload for that week. This is a "learnings in" not "reporting out" type meeting. How those learnings apply to future work is part of the agenda.

This more dynamic process lets something urgent—like a competitive response tool—jump to the front of the line versus just being on a list waiting its turn. It also provides a data-driven forum to discuss what marketing teams should start doing more of or stop doing.

It constantly examines if marketing's outputs are meeting desired outcomes. It also provides a consistent checkpoint on whether or not the product is being portrayed accurately.

It's also where product marketers as strategists play an important role. They check activities against strategies to make sure ideas or campaign assets stay on point.

It does make product marketing quite central to how good marketing gets done. Which begs the question, How do you measure if product marketing is doing a good job? I dive into that next.

Chapter 13
The Metrics That Matter

Unlike sales, which hits its numbers or doesn't, product marketing doesn't have many hard metrics showing a job well done. There are some shorter-term indicators, but for most product marketing outcomes, they take time to know if they worked.

The best ways to measure product marketing come down to what you're trying to get from it. It's why managing expectations about what is wanted from the function is so crucial to do up front.

Product Marketing Objectives

Doing product marketing well is about executing the right range of go-to-market activities necessary to drive the business. That's why product marketing results are so closely tied to company goals.

Because goals vary by company situation and stage, there is no one standard set of measurable objectives or key results. That said, here are some example product marketing OKRs:

- Become recognized category leader by major analyst firm(s)
- Launch product to improve market awareness indicators by 10% and enable adoption by [key new market] as % of customer base

- Position product as redefining [insert concept] as evidenced by inbound or category mentions of [insert concept]
- Enable sales to win >50% of competitive deals
- Grow organic evangelism by 10% on key digital and social platforms

If you're most concerned about enabling revenue, product marketing OKRs will be more aligned with sales objectives. If you're most concerned about market positioning or building awareness, OKRs will focus more on work done together with marketing.

In general, sales and marketing metrics focus on shorter-term business need (like top of funnel size, pipeline, revenue) and product marketing focuses more on what's also important for the longer-term future of the business.

OKRs and KPIs: What's the difference and how do you use them?

OKR (objectives and key results) is a goal-setting framework for defining objectives. They have become the most commonly used framework for fast-moving technology organizations.

Key Results (KR) are specific KPIs with targets. Usually, KRs cover a subset of the KPIs that a team may be tracking. KPI (key performance indicator) is a metric that measures something important. Usually, key results are measured using specific KPIs. Sales cycle time is a KPI. Decreasing sales cycle time by 10% (objective) to 45 days (key result) is an example of an OKR.

When using OKRs, ensure they create alignment across functions. In the case of product marketing, their OKRs must be aligned with product, sales, and marketing teams. Sometimes, OKRs are shared, particularly with marketing teams.

Metrics for Product Marketing

What you do with a metric and what it means depends very much on the lens through which you look. What follows are key metrics—many of them shared with sales, marketing, and product—through the lens of product marketing and some potential actions to improve them.

Product Metrics

- **Happiness, Engagement, Acquisition, Retention, Task Success (HEART) metrics.** Typically owned by product and user experience, there are aspects in adoption and retention that might have product marketing or marketing implications. As the ambassador for the product team, product marketers bring relevant learnings from the product team to marketing teams, wherever it can positively improve a product or user engagement metric. For example, a product losing users during onboarding might have marketing send users a series of emails to get them re-engaged.
- **Customer funnel metrics.** Tracked by marketing or product depending on the GTM model, look for indicators of engagement and flow from one stage to the next. If conversion between stages is lacking, look at what programs, tools, process, or product changes might improve flow through the stages. Product marketers might direct specific tactics—a targeted direct mail, a new call script, video-based training—to improve conversion from one stage to the next.

Marketing Metrics

- **Customer journey engagement.** Typically owned by marketing, they can get data on which content, pages, and websites prospective customers engage with—including third-party websites. Examine this together with sale stage movement over time. Product marketers use this data to help develop what customers or sales needs most to move customers along and work with marketing to make adjustments in the marketing mix.

- **Marketing qualified leads.** Typically owned by marketing, growth in revenue should track with growth in leads or it means something isn't quite working in the customer funnel. In organizations with a product-led motion, product qualified leads (PQLs) either replace or augment how marketing qualified leads gets defined. Product marketers work with marketing teams to help refine target segments, how to engage them, and diagnose what might be in the way increasing leads or predictability.
- **Inbound discovery.** Owned by marketing, inbound organic search on a website plus direct search as a percentage of visitors is one of the indicators of awareness and brand position. The absolute numbers don't tell you much but their shifts over time indicate if more focus on awareness or positioning is required.

Sales Metrics
- **Sales cycle time.** Owned by sales, look for trends that track with the chosen GTM model (e.g. don't expect an enterprise sales cycle to take a few weeks). You want cycles as predictable as possible. When deals go particularly quickly for well-qualified customers, make sure those processes get examined and documented so they can be repeated.
- **Win rates.** This can be a combination of closed won deals versus lost as well as win rates versus the competition. Both metrics are owned by sales, but if either is not ideal for the business, product marketing often takes the lead in examining what's working and what's not. They then take that learning to examine training, tools, process, sales playbook, marketing activity mix, messaging, positioning, pricing, packaging, and third-party evangelism—the full product marketing stack—to see what can have the most positive impact on winning deals.

Financial Metrics

- **Conversion rates by product.** For multi-product companies, monitor if the product mix is aligned with where the business needs to grow. For example, if trying to grow into a new vertical, what percentage of new logos or new revenue comes from that new vertical? If not, ask what incentives, messaging, tools, branding, packaging, pricing, or partnerships might accelerate growth towards the desired mix.

- **Customer acquisition cost (CAC).** This should trend down or stay level if the marketing channel mix, messaging, and positioning are working. If CAC is increasing or unsustainable, examine customer segments, messaging, and the marketing mix.

- **Lifetime customer value (LTV).** If the right customers are being targeted, the lifetime value of a customer versus the cost to acquire it stays at a healthy ratio for the type of business you're in. If this ratio isn't good, it can indicate product issues (not enough value to stay a customer), positioning (expectations different from what product delivered), or even sales training issues (what customer was sold versus what they expected was different). Examining this together with other teams is a collective endeavor, but inevitably has product marketing implications.

- **Retention.** An important but lagging indicator if customers are getting enough perceived value to stay a customer. Net Promoter Scores or other periodic customer satisfaction metrics can be leading indicators for retention. If retention isn't at its desired rate (this varies by business type), product marketing examines all go-to-market materials to see if there are disconnects with customers' expectations, sales process, price, or packaging issues.

People and Process

Practice Patience and Persistence

Good product marketing takes time, and measuring its effectiveness does too. Like all teams, product marketing does best when it knows how its success will be measured.

This requires clarity on what a company expects from the role. In turn, product marketers use this to respond to market realities with appropriate go-to-market actions.

Part 3 is where I go into great depth on different levers and concepts that shape strong product go-to-markets. Done well, they set a frame from which product and GTM teams have clarity of purpose and make decisions that serve the business well.

People and Process

Part Three

Strategist:
Guardrails and Levers:
The Tools to Guide Product
Go-to-Market Strategy

Chapter 14

When Strategy Guides Product Go-to-Market

Salesforce

Dreamforce is like the Super Bowl for SaaS. Salesforce's annual event in San Francisco brings together over 170,000 people from over 120 countries. Dozens of products get launched amid thousands of sessions with speakers as wide-ranging as Olympic champions, Michelin-star chefs, and titans of industry. It's the kind of seminal event that can instantly catapult products into massive market awareness.

But Michelle Jones decided she shouldn't launch her product there.

As the then director of product marketing for Salesforce's B2B Commerce software, she decided a much smaller Salesforce Connections conference in Chicago earlier in the year was better for her product. This decision, together with a series of many other thoughtful go-to-market actions, is how Michelle and her team propelled B2B Commerce to become one of the major drivers of growth for Salesforce at the time.

Confident and determined, Michelle began her career as an analyst at Ernst & Young and merchandise planner at the Gap. It meant she was good at understanding customers and markets. B2B Commerce's customers were companies trying to sell goods and for whom the fourth quarter ramp to the holiday

113

buying season was their busiest. They would have no time for anything new in Dreamforce's usual fall timeframe.

Picking a launch event that worked for her product's timeline was the easy part. Michelle had to make her product worthy—among Salesforce's more than 90 others at the time—to get featured in a center stage keynote and be ready for Salesforce's massive sales team to sell.

She created a story, marrying market trends with customer benefits, showing how the Salesforce platform provided a powerful engine for business growth. Customers could easily add functionality that let them add the smooth buying experiences people saw on B2C e-commerce sites to a B2B transaction and improve order accuracy. It let them get more out of an investment they had already made.

She knew the story's credibility would be more powerful if supported by customers already experiencing those benefits. She made sure by launch three very different customers were willing to be public examples: industrial products manufacturer Ecolab, chemical distributor Univar, and consumer brand manufacturer L'Oréal.

Gartner, as an analyst name brand, carried disproportionate influence. So Michelle worked with the product and analyst relations teams to secure the early backing of Penny Gillespie, then a research vice president and analyst at Gartner. They were able to quote her in the product's launch press release saying, "This covers a big gap in functionality for Salesforce when it comes to digital commerce. It also completes the B2B channel for Salesforce, complementing its existing applications including sales enablement, procure-to-pay and order-entry portals."

But the most important evangelists Michelle had to enable were her internal sales teams. The biggest market for her product was existing Salesforce customers, so the sales tools she created had to start with that in mind. She included the broader platform story and its benefits in addition to the product's specific feature benefits—like easy reorders or custom catalogs. She made sure sales incentives were in place and that she had packaged up all customer and analyst validation to make sales' jobs easy from day one all the way through the second half of the year when Dreamforce took so many more cycles.

All this is why her product did so well in its debut year.

A go-to-market like B2B Commerce might look obvious when we read about it now. But Michelle didn't always do the obvious: she didn't launch at Dreamforce, she prioritized the internal work (versus the external) to make her product worth sales' effort and a major executive keynote, and she created analyst relationships before the product even launched. She didn't do all the work by herself, but she made sure it all happened.

This is what a strategic and comprehensive approach to product go-to-market looks like. Great product marketers, like Michelle, seamlessly weave together the four fundamentals in how they practice the craft.

- Fundamental 1: Ambassador. She understood her customers well enough to align her biggest marketing actions with timing that worked best for them and the product. She also realized her internal salesforce was an equally important customer.
- Fundamental 2: Strategist. Instead of going with the biggest opportunity to launch (Dreamforce) she went with what was smartest for her product. She intentionally tied her product to the benefits of the larger Salesforce ecosystem.
- Fundamental 3: Storyteller. She wove a story in which her product had meaning—for customers, the industry, and for Salesforce—all of which was important in standing out.
- Fundamental 4: Evangelist. She knew which analysts, customers, and sales teams could have the biggest impact for her product and made sure all were ready at launch.

Strategist

In too many companies, product marketing is about the tasks and not the role's bigger purpose: driving product adoption by shaping market perception through strategic marketing activities that meet business goals. Every action should build toward an intentional outcome. It's how you get superior results.

A checklist version of what Michelle did—product messaging, press release, customer and analyst testimonials, sales first-call deck, launch presentation, sales playbook—removes all the context that made hers a really good product go-to-market, in particular the strategic context that shaped *when* and *why*.

This section of the book goes deep on the tools and techniques that make product go-to-market more strategic and impactful. I explore what strategic guardrails do when in place right at the start and how to rethink often misunderstood concepts, like the technology adoption curve, brand, pricing, and integrated campaigns. I explain how it all comes together in a lightweight, canvas-style framework that can improve organizational alignment for any product go-to-market. This section concludes with a chapter on how product go-to-markets show up in larger marketing plans so you can recognize how they intersect.

Strategist

Chapter 15

What the iPhone Shows Us About Adoption Life Cycles

I once worked with an online backup company whose initial target customers were "Anyone whose money was green." They discovered paid radio host endorsements (a more modern equivalent is podcasts) worked particularly well for capturing customers. But as often happens with paid customer acquisition, it was becoming too expensive to rely on to grow.

This made the marketing team do a deeper dive on their customer base. They were shocked to learn it skewed old—senior citizen old—in retrospect, not surprising given their heavy reliance on radio.

These customers were easy to retain but did nothing to help the business grow. There was no word of mouth—people didn't tend to ask their parents for online backup advice—and they weren't the sort that wrote reviews.

The company had to completely overhaul their marketing activities and focus more on channels that brought in customers who talked about products *with others*. This wasn't just important to make the math of the business work; it was an important signal around true market fit. If they had it, they'd have more organic growth.

They looked again to their customer analysis and found their product's sweet spot was not 20-somethings who spent a lot of time on social media. Rather, it was people in their early

30s who listened to NPR (National Public Radio) and felt they had more to lose—family photos or an archive of work.

Go-to-market shifted from gaining any customer to getting the right customers that let them grow sustainably over time. It was an expensive and time-consuming lesson in the importance of paying attention to the technology adoption curve, and it is one of the biggest misses I consistently see in how products, especially for startups, go to market.

Let's Talk Technology Adoption Life Cycle

The technology adoption life cycle (aka innovation adoption life cycle or technology adoption curve) is how any new technology product flows through different adopter groups. It follows a basic bell curve with first adopters (Figure 15.1) called "innovators," followed by "early adopters." The fat part of the curve is "early majority" and "late majority" before it contracts down for the "laggards," who are slow to adopt any new technology.

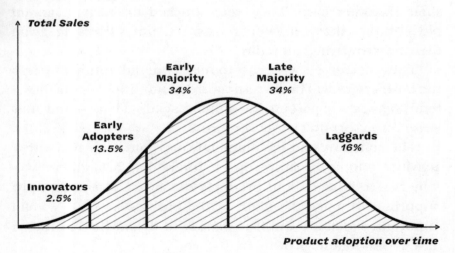

Figure 15.1   The classic technology adoption curve.

These adoption cycles typically take 7 to 10 or more years to move through. People can only adopt technology so fast. Marketing awareness and activities must build and layer over time.

The mistakes I see most often are when teams believe they're moving through the adopter segments faster than they really are, or they don't realize their initial market segments don't set them up well for the next stage of growth.

This is hard stuff because there is no easy way to know where a product is on this curve. Movement through it isn't evenly distributed over time. A product can spend years at the earlier stages of the adoption curve and move more rapidly through the "majority" stage—as often happens for enterprise companies—or vice versa, as is the case for many "hot" consumer tech companies.

Add to this that customer segments are not just demographic, psychographic, or industry verticals. In modern marketing, technographics (technologies used), firmographics (size, geography, industry), behaviors (likelihood to buy), intent (looking at competitive websites, reading content online), product usage (features most used by customers who renew), browsing history, and the actions of people inside a company (referred to as "account-based" actions) all shape how marketers segment or target a market.

From product marketing's perspective, applying the curve to go-to-market thinking means understanding the impact each customer segment has on subsequent customers' product adoption. The online backup company accidentally having seniors as early adopters was a great example. Despite what looked like good product adoption, they didn't have a customer base from which to cost-effectively grow (Figure 15.2).

Fixing this was not about the product; it took a lot of marketing adjustments and time.

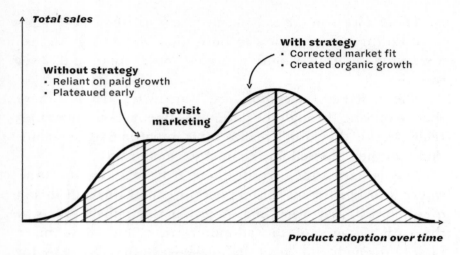

Figure 15.2   Growth stalled because existing customers weren't the best ones from which the business could continue to grow without it costing too much.

Product managers, in contrast, are always looking for the next customers with new needs to solve. This means they move more quickly through their definition of customer segments, using jobs-to-be-done or problems-to-solve to define them, not addressable customers in market who are likely to buy. It's how product can get ahead of where go-to-market teams feel the market is.

Product marketers help bridge this gap. As the ambassador to product teams, adoption curve thinking should inform product decisions. As the ambassador to the marketing team, they should help prioritize customer segments so marketing teams focus on getting the right customers. As storytellers and evangelists, product marketers shape categories, so a product can be seen as providing value, even when it's ahead of the market.

For marketing teams, how customers are segmented varies wildly. A startup might stick with four verticals over multiple releases over 18 months, while a more mature company adds two new verticals a month with no updates in product.

After teams identify which customer segments are most meaningful in *shaping market adoption*, the plan on how to target them through marketing strategies and actions is what goes into a product's go-to-market plan.

Applying Life-Cycle Dynamics to Product Go-to-Market Actions

Many startups don't know who their best initial target customers are. This is normal and the iterative work of product/market fit is what helps find them.

Superhuman's Rahul Vohra detailed his team's journey finding theirs.[1] They defined the most discerning person among their early customers as "the one[s] who would enjoy the product for its greatest benefit and help spread the word." This drove their product strategy, but just as important, it shaped their product go-to-market.

Some of what they did included:

- An extremely clear, differentiated market position: "the fastest email experience ever made"
- For early adopters, the ability to refer friends to "jump to the front of the beta line" (which became tens of thousands long)
- Getting qualified before being accepted into the beta (you couldn't be a user if they didn't think you'd experience the product's benefits)
- Monetization at the start of onboarding (no free trial)
- A required onboarding call to ensure customers could be successful right from the start

---

[1] Rahul Vohra, "How Superhuman Built an Engine to Find Product Market Fit," First Round Review, n.d., https://review.firstround.com/how-super human-built-an-engine-to-find-product-market-fit.

These were counterintuitive go-to-market practices at the time—free trial, unlimited betas, and self-serve onboarding were the industry norms. But they did them with confidence because they had so much clarity on their best early adopters to drive future success. They said "sorry, not yet" to any early customer they weren't sure would help them succeed.

Let's look at how this curve and product go-to-market actions might play out in a B2B scenario, where the economic buyer and decision-maker differs from the user feeling the daily pain.

In the cybersecurity market, the chief information security officer (CISO) persona is the most sought-after senior decision-maker. Palo Alto Networks, a publicly traded company, can leverage brand awareness, existing products, CISO programs, and a large sales team in their go-to-market. Their ability to move through this curve will be very different from a lesser-known startup.

In a startup's case, their focus might be on who experiences the daily pain their product solves. Product go-to-market activities must focus on what activates their users to influence the CISO. Same end decision-maker, but very different approaches to market because the companies have different strengths. Which specific customer segments and marketing activities to target for growth differ, even with the same end decision-maker.

Watch the tendency to overgeneralize customer segments when doing product go-to-market planning. One I see all the time: SMB, mid-market, enterprise. When it comes to designing a go-to-market, a mid-market company is not the customer. A "mid-level supply-chain manager who is willing to adopt new technology and directs purchases across at least 50 vendors" could be. Likewise, for consumer-facing companies, a marketable customer segment is not "everyone who uses the Internet"; it's "a word-lover who is curious about new words and language."

Be Thoughtful and Patient

Don't underestimate the work it takes to move through the adoption life cycle and how important it is to do in a deliberate way. Even the iPhone, possibly the most successful technology product of all time, took considered market actions over years to shift buying behavior.

I'll close this chapter with a walk-through of how key elements in their product go-to-market worked together.

Hello iPhone: The Technology Adoption Curve in Action

I'm going to massively oversimplify the many things Apple did, but here's what the iPhone looked like as it went through the adoption curve segments in my personal circle. Figure 15.3 (on page 126) shows the key activities along the adoption curve in context.

Innovators
My Apple fanatic friend, Mike, stood in line to get the first iPhone on day 1. He did a mass email to all his friends with his first impressions days later:

> It is every bit as fabulous as expected. Just stunning. Videos are great and if you don't have headphones plugged in, it uses the external speaker, making it really nice to show videos to your friend.
> The keyboard is amazingly easy! Perfect for the "mature" geek.
> :-)
> P.S. I got 2 in case I lose one.

(continued)

Strategist

The price was a then-astounding $599, priced to encourage slower adoption by a more exclusive group with the resources and behaviors that fan the flames of future adopters.

Early adopters

Just months later, the same phone with an OS upgrade dropped its price to $499. With Mike's endorsement, I felt comfortable getting one, as it was now at a price I was more comfortable spending. Most of my techie friends not in the first wave did the same.

Early majority

A year later, Apple launched the iPhone 3G with a multiyear exclusive distribution deal with AT&T. This let them drop the price to $199. It is also when they launched the App Store. Over 10 million apps[2] were downloaded in the first weekend. This is when my siblings—who work in finance, the military, and commercial production—got their iPhones. The price made it more accessible to them, and it was a highly sought after gift or personal splurge.

Two years later, when the iPhone 4 was launched on Verizon's network with Retina Display, the first front-facing camera, and FaceTime, my mother-in-law got one. By then, the iPhone was being used by everyone she trusted, including all her grandkids, so she was ready to make the leap.

Late majority and laggards

This was in the iPhone 5/6/7 timeframe, roughly seven years into the iPhone's life-span. It's the era when my mother—a German immigrant and the most resistant technology adopter I know—joined the iPhone bandwagon. It had nothing to do with the technology. We foisted it on her for Mother's Day. She agreed it was a more convenient digital camera and liked FaceTime to see her grandchildren. These were all very old features at that point but they were the ones that made her adopt it.

---

[2] Apple, "iPhone App Store Downloads Top 10 Million in First Weekend," press release, July 14, 2008, https://www.apple.com/newsroom/2008/07/14iPhone-App-Store-Downloads-Top-10-Million-in-First-Weekend/.

Strategist

What can you learn?

Even if Apple's long-term goal was adoption by people like my mother, they didn't start there. They started with early adopters like Mike. Getting to my mom took the better part of a decade and a steady evolution in product, distribution, pricing, advertising, and promotions. Ten years in, with the iPhone X vanquishing the once iconic "home" button and a growing adjacent product ecosystem in the Apple Watch, they were on to their next adoption curve. If the iPhone is one of the most successful products of all time—with a marketing budget to match—think about how long it will take *your* product to get through its adoption life cycle.

(continued)

Strategist

Strategist

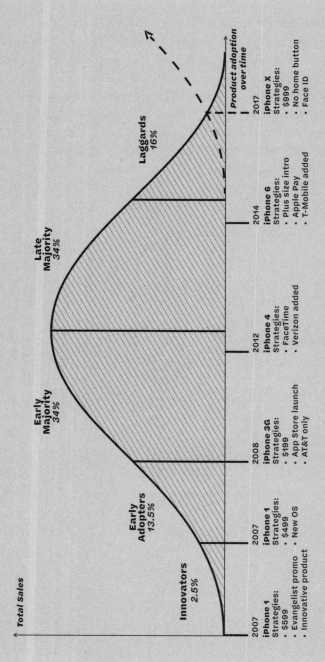

Figure 15.3 Well-planned marketing strategies and tactics combined with strong product innovations propelled the iPhone through its adoption curve.

Chapter 16

The Brand Lever

It's Not What You Think

Netflix didn't become what it is today because they did a Super Bowl ad or blanketed cities with billboards. They built their brand by encouraging people to try Netflix, then delivered a consistently great experience—whether it was red envelopes arriving quickly or bingeworthy entertainment easy to watch anytime, anywhere.

As a company, Netflix famously has very empowered product teams. They are expected to make the right decisions for the business. This was put to the test when the team in charge of the homepage converting people from a free trial faced a decision that could cost Netflix tens of millions in lost revenue.

When they notified customers before their credit card got charged at the end of their 30-day trial, more people ended their subscriptions who might have otherwise just forgotten to cancel.

The team had to choose: more respectful, proactive notice that maintains a positive brand impression or get tens of millions in extra revenue each month.

They decided Netflix's brand reputation was the right choice for future growth. And they were right; that decision was tens of millions of global subscribers ago.[1]

Brand isn't done well by most tech companies because it is largely misunderstood. It is presumed to be a company's name, logo, colors, design, and tone of voice—all things Netflix also does very well. Brand includes all this, but it is more about a consistent experience across every aspect of how a company acts—it's a promise between a customer and a company.

The best brands in the world deliver their promise in every part of a customer's experience—even something as perfunctory as notifying of a charge about to hit an account.

In part 1 of the book, I talk about how product landscapes are more crowded than ever, making differentiation hard to achieve. Brand is a tool for strategic advantage and should be used that way. But it requires careful attention to a surprising number of little things.

Much of brand management and execution is beyond the domain of product marketing. But there are situations where it lives at the product level, including: product or feature naming, moving toward a multiproduct suite, moving to a line of business brand, or diagnosing when there is a company brand issue, not a product one.

This chapter focuses on better understanding brand basics and the circumstances where brand work at the product level can improve a product's go-to-market outcomes.

Brand in Tech

When WebFilings started back in 2008, no one would call their logo, brand colors, or company name particularly inspired. But customers *loved* them. They had first-name access to a dedicated customer success manager who offered suggestions on how to use the product better at weekly check-ins. They

---

[1] I highly recommend Gibson Biddle's talks on product strategy and leadership, which is where this story came from.

made customers feel like they couldn't fail. It built unmatched loyalty for the brand propelling them to category leadership. They are now the public company, Workiva. They had a great product, of course, but the loyalty people had to the brand came from an early and deep investment in customer success.

For most people, the brand of a tech company is driven primarily by their product experience. But for a customer, this also includes their experience across product support, the sales process, and even pricing. This is in addition to the more obvious expressions of brand such as a website, social media, or advertising.

In highly competitive, crowded markets, your brand experience can make a product *feel* different. It's a way to shift the playing field beyond features and punch above your weight for earlier stage companies. Attention to brand signals maturity.

It is also why brand management tends to be driven at the company-wide marketing level. Products change; brand promises don't. Products end-of-life, more get added, and functionality can change a lot. When a company invests well in its corporate brand, customer loyalty lives across products or despite them. That's the end goal of brand for growing, multiproduct companies: long-term loyalty.

When product marketers plan a product's go-to-market, they should consider their company's overall brand situation as part of the strategy mix. Does it need to be improved or reframed? How does the product deliver against the desired brand experience?

What follows are common brand situations faced by product marketing. Brand as a discipline is deep, so consider this just an introduction to concepts to get you thinking about them more strategically.

Product Scope Expands; Market Perception of Current Product Is Narrow

This was what happened with Pocket back in chapter 1. Read It Later, as a product name, no longer encompassed what they did—they now saved videos, images, and shopping links. They also needed to show the world they were more than just a feature or app.

Pocket, as a name, could encompass all their new functionality. It also worked for Pocket Hits, their weekly email, seen by more people each week than used their mobile app. The rebrand of the product worked so well, it became the company's name.

Brand Strategy Informs Product Go-to-Market Strategy

Any multiproduct tech company must have a broader brand strategy. There are three brand levels to pay attention to: company (e.g. Apple), line of business or key brand (e.g. Music, TV, Watch, iPhone, Mac), and products (e.g. MacBook Air, MacBook Pro, iMac)

These different levels exist to simplify how customers understand a company. It creates a categorization system for customers, but it does *not* mean each level gets marketed. For example, Apple markets at the product level—Apple TV+, the newest MacBook or iPhone—and not at the line of business level. But if you're trying to find the latest iPhone options, you'd look to the iPhone section of the website, then look at all the product options.

Microsoft, conversely, tends to market a lot at the line of business or product suite level: Office, Xbox, Azure.

Larger company brand strategy often determines how an individual product gets marketed. A product's go-to-market can be driven by the importance of a line-of-business to a company. For example, when I was on the Office team, Word's marketing was reduced to a small number of features that supported marketing Office as an integrated productivity suite.

Leverage Existing Product-Brand Loyalty to Acquire New Audiences

Intuit's product brands—QuickBooks and TurboTax—are better known to their target audiences than the company brand.

Because of that, product brands are used as a shortcut to signal who a product is for and what it does. Looking for small business accounting solutions? They're easy to find: QuickBooks, QuickBooks Payroll, QuickBooks Time.

Another example is Atlassian's Jira Software—a brand dominant among developers. Atlassian uses the Jira brand in other products, such as Jira Service Management and Jira Align to lend Jira's developer credibility to those products, even though they don't target the exact same audiences.

Naming decisions at this level are often driven by product marketing. It's crucial to pay attention to where brand loyalty lives and use it with intention.

Improve Ability to Penetrate a Market with a New Brand

Microsoft as a brand is firmly associated with enterprise and productivity software, not edgy gaming. That's why it created the Xbox brand, which it now uses for all its products and marketing to the gamer audience.

Sometimes, companies need to create a different line of business or key brand because the existing brand has too much baggage that doesn't connect with a target market. Marketing at a key brand or line of business brand level is a decision often driven by corporate brand decision-makers. But its opportunities and limitations shape a product's go-to-market.

Another version of this is unifying a group of integrated or connected products under a product suite name. Especially when a company has multiple products targeting the same audience, it vastly simplifies go-to-market to unite them under one umbrella versus marketing each individually.

Product Naming Is Brand Strategy

How things are named helps people understand what a product does and for whom. For example, in Salesforce's current Marketing Cloud, products include Email Studio, Audience Studio,

Strategist

Mobile Studio, Social Studio, and Data Studio, among others. Even if you don't know the products, you do know if Studio is in the name, it is part of the Marketing Cloud. Consistent prefixes and suffixes are a popular way to connect products that live in a broader product family.

Product marketers drive this naming thinking because they are expected to have visibility into product roadmaps. Naming taxonomies affect future product naming. They also play a vital role in identifying if a bigger aspect of brand strategy affects product go-to-market.

Sometimes There Is a Bigger Battlefront

When I was at Netscape Communications, the company was the poster child of all the new possibilities of the Internet era. It was the original high-flying company of the Internet. Most people called our market dominant browser "Netscape" even though the product name was Netscape Navigator.

No one worried about it when everything was going well. This was despite an increasing part of the company's revenue coming out of other divisions, like the server division.

But when Microsoft's full-force attack with Internet Explorer made massive inroads on the browser side, the perception was "as goes the browser, so goes Netscape" (i.e. embattled). It didn't matter that our growth businesses were going strong or that we were innovating new Internet protocols and services.

We had a company brand problem, but we kept fighting it with product wars. It was the product marketing team going through a product naming exercise that flagged the bigger company brand issue. But at the corporate marketing level, they still wanted to use the product loyalty associated with Netscape to increase awareness of other products. The company was eventually bought by AOL, largely because Netscape's brand was associated with a market segment the AOL brand couldn't capture. It seared into me the importance of constant examination of the relationship between products and their larger company brands.

Product naming and brand strategy are often an afterthought at many tech companies. But brand strategy matters—it's how companies achieve more than the sum of their individual parts.

Good product marketing makes sure product-brand strategy is done thoughtfully. Great execution with brand can shape entire categories.

Chapter 17

The Pricing Lever

It's About Perceived Value

Nike used to host a women's marathon in my hometown of San Francisco that brought 10,000 women together for a running love-fest. From Olympians to hairstyling stations at registration to tents with the latest Nike gear, the festival atmosphere celebrated women and running. The race also served a valuable secondary purpose: it was a living, breathing market research petri dish for Nike.

It's how I found myself in a focus group for Nike sunglasses. Athletic and energetic to a person, we free ranged around a room that was like a sunglasses counter without the glass: we could see, touch, and try on everything.

After a period of time, we sat down and filled out a stack ranking of our top choices using whatever mattered most to us personally—sportiness, trendiness, feel, color.

And then they revealed the prices.

I'm a very pragmatic person and don't buy sunglasses often. When I saw the pair I'd ranked as number one cost $300, it immediately dropped to my lowest choice. That was more than I was willing to pay for fashion sunglasses from Nike. But I was willing to pay $125 for a performance pair designed for racing. Those formerly-ranked-number-four sunglasses were now my number one pick. The perceived value matched my desired use case.

This is the simple calculus for everyone with pricing: if I like this product, is it worth it to me at this price from this company? Obviously brand plays an enormous role in this equation. It's why Apple's watches price from $199 to $399 while Google's Fitbit watches price in the $179 to $299 range despite having similar features.

Pricing is not about what it costs to produce a service. It's about people's perceived value of a product and their willingness to pay.

Pricing is value engineering.

Pricing Basics

It's important to look at pricing from the lens of the customer: how much do they value what you do for them relative to the price they're paying *your company*. Price is not an absolute value; it's a relative one.

Often when people think their product has a pricing problem, the real problem is the perceived value isn't there. Price was set relative to similar products in the same category, not to market sentiment or brand perceptions.

Context (how pricing is presented on a page, especially relative to other options), brand, competitive alternatives, availability, ease, budget—all this and more factor into a customer's value calculous.

Combine this with the fact that modern products are built on technical infrastructures that are extremely complex and dynamic (e.g. variable cloud and compute costs), and it's clear why pricing has become a highly specialized practice.

Because of that, this chapter is intended as just a primer on key concepts. Let's start with who owns pricing.

It varies wildly based on go-to-market, business models, product complexity, leadership, and company stage. While it increasingly falls to specialists, I've seen just as many product managers own it as I have product marketers. I've heard equally strong arguments on why each should (or shouldn't). They are both critical influencers, but let's separate key aspects of pricing to help you decide who should own it at your company.

- **Monetization strategy**—deciding how and when you'll make money—is distinct from how much you'll charge. It is often a joint exercise between some mix of product, business operations, and product marketing, and it is largely dependent on the range of skills in the people doing those roles.
- **Pricing strategy**—determining the actual price for products—is often heavily driven or influenced by teams in finance or business operations. For more mature companies, pricing is shaped by customer funnel dynamics, conversion estimates, scenario forecasting, and the cost to build the product and support each customer. For anything sold by a salesforce, they provide input. Framing the perception of value for the customer falls to product marketing.
- **Packaging strategy** is distinct from pricing. Packaging is about creating bundles that serve customers, market segments, or use cases in ways that support a business. It's important not to conflate pricing and packaging; they do different things. Packaging is typically driven by product marketing with product a big partner in enabling it.

Wherever pricing and packaging lives, the product marketer's role is to ground it in real value for customers and make sure it serves different customer segments and business goals well.

The rest of this chapter provides additional guiding principles for how to create smart pricing and packaging.

Make Pricing Easy for Customers and Good for Business

Pricing should be built on a unit of measure that is easily understood by customers and connects to your company's financial success. Easier said than done. But when selling tech products, you want the hard thinking to be around whether or not to buy your product, not understanding how it's priced.

Here are four essential concepts when thinking this through:

1. **Use a metric that reflects the value of your product and grows as your product provides more value.** For example, Dropbox costs more with each additional terabyte of storage required. Want more? Pay more. Or with a managed service, as the size of the attack surface increases—the larger your organization and higher the complexity—the more the platform service costs. Keep it simple.

2. **It must be simple enough for customers to do the math in their heads.** People like to know what they're in for so they can assess if the value is worth the cost. Per user pricing works because it's easy to calculate what the cost might be depending on how many people are on a particular team. If you're an infrastructure service whose price depends on usage, you might think pay-per-use seems attractive because people won't overpay. Consider if you should pre-package when customers reach a particular usage threshold. You want to encourage your customers to use your product as much as possible, so your product stays sticky.

3. **It must be easily measurable.** The IT and/or compliance teams have to know whether or not they are complying with the terms of a contract. For example, per seat pricing seems simple on the surface, but if you're trying to grow usage inside a company rapidly, individual licenses

Strategist

can be an impediment. To control costs, IT tries to control the number of licenses. But if anyone can use your product, and it's just paid for based on usage, then there is nothing slowing its spread. IT and compliance teams like being able to predict costs, so pay attention to essential #2 and #4.

4. **It must be something a CFO or procurement person will understand as they compare your cost to others.** They won't understand the nuances of your differentiators. They're coming at this from a total cost of ownership and return on investment point of view. Does your product or service seem to be worth the cost? This isn't just relative to your direct competition. They're comparing your perceived worth relative to other contracts.

What Drives Your Business?

When it comes to pricing, the psychology can be oversimplified into two types: those who buy because it's cheap and those who buy because it's the best. Decide which end of the pricing spectrum best matches your product and go-to-market.

For example, you might price inexpensively or have a free version to capture customers that could be evangelists for your product. But the flip side is equally true. In chapter 15, we saw the iPhone's initial premium pricing was about keeping beachhead customers small in number and feeling exclusive.

Whether or not you can be a premium product is about perception and brand. Is this product worth the price from your company? My reaction to $300 fashion sunglasses from Nike was very different than it might have been if they were from Gucci.

If you decide you're a premium product because your market supports that strategy, if you *don't* hear customer complaints that you're expensive, you're leaving money on the table.

I once met with a very-early-stage startup sales leader who was trying to understand how to price their product. He said, "I basically double the price at every meeting and am waiting for someone to blink. So far, that hasn't happened. We still don't know its price."

You can't be sure about your products' value until you're out there in the market, pricing with actual customers. Look for breaking points.

If you're an early-stage B2B company, you might not know until you hit 20 to 30 customers. It's okay if prior to that pricing is customized—you're figuring out what will scale. After that threshold, you want to establish repeatable guardrails.

If you're a B2C company, you might frequently revisit pricing because you distribute online and price testing is easy.

Use Packaging for Customer Segments or Use Cases

The job of packaging is to move people toward and not away from a purchase decision. It's counterintuitive, but lots of options attract gawkers and can prevent purchases.

Decision fatigue is a real thing. Complexity in pricing and packaging creates friction in the sales and buying process. Table 17.1 provides an overview of packaging options and how to think about them.

Table 17.1   A guide to when and why for different packaging strategies

Packaging Strategies	All Inclusive	Platform + Add-Ons	Good, Better, Best
When	Early product	Expanding product line(s) and feature sets	Well-defined use cases and market segments
Why to do it	Simple for customers Simple for sales	Flexible for customers but brings complexity Lets sales customize approach	Drive highest possible ASP Less decision fatigue Repeatable sales execution

One of the most common ways to package is with editions, which prepackage for customer segments or use cases while also driving toward what's important for your business. A general guideline on editions in the entry-level should include the must-have product(s). You want the vast majority of potential customers to think "look at all I get for this price."

The next-level edition might contain must-haves for specific market or customer segments. It's not for everyone, but for those who need it, it is high value or clearly differentiated. This may also apply for any additional editions or add-ons. They are niche, intended for highly specific use cases.

Trends Shift Expectations of Value

Purchasing behavior has shifted toward cloud and subscription in both D2C (think Netflix, Spotify, Apple Music/TV+) and B2B (Google or Salesforce). If you'd bought a Microsoft Office Home & Student edition, it used to be $104.99, and you might use the same version for years. The now-renamed Microsoft 365 gives you more and costs $99.99 a year.

Like anything, pricing and packaging shifts with trends and consumer expectations. Smart pricing and packaging can remove unnecessary friction and enhance the perception of value. But most importantly, it is one of your most powerful business levers.

Strategist

Profile in Product Marketing: Jenn Wei

Jenn Wei spent her career in product management and product marketing positions at VMWare, DocuSign, and currently as the Vice President of Product Growth at Rubrik. She embodies the excellence I consistently see in careers combining both product management and product marketing. She is also one of the clearest thinkers on pricing and packaging I know. I owe a lot in this chapter to Jenn who shared her thinking with me.

Chapter 18

Marketing When It's Not About Product

I'm writing this at a time when the world is emerging from a global pandemic. Some sectors are booming, others struggling. Work from home is the new normal and "offices" will never be the same.

$#*! happens. It affects us all. It's why the forces of nature, the steady march of life, and world events *should* have a massive impact on how we market. And they have nothing to do with your product.

Whether driven by market, technical, or resource constraints, all companies experience times when they must maintain marketing momentum without anything significant from product. This feels scary because in tech, we're used to focusing on new being what sells.

For product marketing it means rethinking how to market when product can't be used as the marketing catalyst. It's actually a wonderful time to work with marketing to double-down on things that are important to improve.

Nothing in this chapter is the sole domain of product marketing. But because product marketing is constantly interpreting market signals, whether it's by contributing content, programs, or squeezing leverage out of old product, product marketing plays a key role in marketing success when it's not about the product.

Campaigns Beyond Product

In the world of marketing, a campaign is a specific collection of coordinated actions to address a specific market opportunity or challenge. Marketing is constantly running campaigns. It's a good practice to not make them all about product. Focus instead on what specific audiences care about or company-level momentum.

Some examples worthy of coordinated efforts:

- Leverage a black-swan event (e.g. global pandemic—see the example in the callout at the end of this chapter)
- Target a particular micro vertical (e.g. sole practitioner accounting firms)
- Amplify a singular company event (e.g. acquisition or going public)
- Activate a dormant or existing customer segment (e.g. emails that haven't been marketed to in a while)
- Shift brand perception (e.g. brand isn't viewed as innovative anymore)
- Convert users of a competitive product (e.g. dedicated program making it easy for users to switch)

The what and how usually falls to marketing, but architecting the *why* and *when* for a campaign might come from product marketing if they see a unique opportunity or challenge that can be addressed in a more targeted go-to-market approach.

Invest in the Emotionality of Your Brand

I went into depth on brand in chapter 16, but this is a great time to emphasize the emotional over the rational. Brand is about what people aspire to be—a good mom, a capable leader, an innovator, eternally young—and the best ones make us feel good about ourselves.

The emotional aspects of brand are often underinvested in by technology companies because it doesn't directly correlate to sales. This is where small actions can add up to a big difference. How can marketing programs deepen *relationships* with customers?

Improve Collaboration Between Marketing and Sales

It's easy to believe you already have this in place. Weekly meetings and communication exist. Target account lists are jointly developed. Campaigns are run that feed the pipeline.

But there is always room to take this collaboration to a deeper level. For example, creating tightly coordinated campaigns for a moment in time. Choosing to be hyper-focused on just one vertical or creating more customized, much smaller campaigns or activities.

Because product marketers are deeply aware of the market, providing input and inspiration to both sales and marketing teams is key. Nothing new in product is a great time to push for marketing creativity and sales allyship.

Examine the Customer Journey

In the early days of Peloton, they famously focused on in-mall pods for their initial sales. This let prospective buyers use the product and experience the most expensive part—the bike—and how it was different.

But just as important: it let Peloton engage in direct conversations with customers and understand their target market. They could hear and see what converted casual interest to a customer. They didn't use digital channels heavily until they had more confidence in what made people buy.

So much of people's consideration of a product happens in a blended way—dominated by online but offline too.

Strategist

People have conversations with others they know. Often, people might not even know they're in the market until they see or experience something.

Periods without major product enhancements are a great time to pay closer attention to points of influence. Take the time to reach out to customers with high NPS scores and ask them if they'd be willing to write reviews on a comparison site or do a video testimonial. Or just send them a thank-you gift to delight them.

Consider revisiting the marketing mix and whether traditional marketing (outdoor, radio, TV) might work for growth. If marketing is a long game with customers over time, mix up how you play the game.

Enable Evangelism from Customers

There is nothing more powerful than your customers advocating on your behalf. You want wildly evangelistic fans that *want* to tell the world about your product because they truly believe others should know how great it is.

For whatever reason, enabling evangelism from existing customers often maps to customer referral programs on a lot of teams. Do better.

Marketing to customers begins by making them successful using your product. These might be actions the customer success team takes, with marketing sending a follow-up "We appreciate you!" thank-you package.

Customers want to feel like they're in a trusted relationship with companies. This is especially true if they have ongoing, expensive contracts. What can you do to make them really like you and not just think of you as a technology vendor?

Strategist

Activate Your Community

Community means many things, but at its core, it's about scaling evangelism. There is only so much a company can do directly. And it's always more effective if someone else does it on your behalf.

Community can range from providing a forum for users to ask questions, to geographic meetups, to having an army of experts who can help people problem-solve on their own.

It can also be something like a customer council where learning is two-way—customers give in-depth feedback on what can be better, and you likewise learn how you can better serve them. This tactic in particular is often driven by product marketing together with product management.

The mistake most make with community is they focus on tools or events—the scaffolding—and then assume community is in place. Keep it simple: can loyal users who feel proud to be associated with your brand discuss it in a public way together with others?

If you're good at creating actual community—not just putting in place the scaffolding—social media is a great place to measure it.

Strategist

How Modern Hire Acted Swiftly to Create a Customer-Centric Coordinated Campaign

Like so many businesses, Modern Hire shut down their offices in mid-March of 2020 and went to 100% remote work. Rather than pull back from all marketing, Jay Miller, their head of marketing who came from product marketing, had his team of eight take two weeks to strategize and find new opportunities from the shutdown.

(continued)

Modern Hire is an HR Tech company whose product combines AI and psychological science to improve every aspect of hiring workflow and candidate experience. It was not an obvious need as many businesses immediately contracted their workforces.

Nearly all of their competitors went to "free video interviewing" overnight. Jay's team decided the best play was to elevate the importance of better candidate experiences and smarter hiring decisions. It made sense because everyone had to shift to remote interviewing and hiring instantly, and data augmenting those decisions would be more valued.

They crafted a comprehensive "Let's get to work" campaign in just two weeks. It launched iteratively as elements became ready.

This is what they did with their sales team before the campaign was rolled out:

- Educated them on messaging: tone and differentiating specifics
- Provided email templates, scripts, and new talking points that addressed the new normal with empathy and constructive ways to be genuinely helpful
- Competitive sell sheet contrasting using Modern Hire as a platform approach versus just using conferencing tools (Zoom, Teams, or Google Meet) as well as a customer facing brochure
- Prepped on client case studies
- Updated partnership slides

The marketing campaign itself included

- A home page refresh
- COVID landing page
- Expanded product and industry pages
- Weekly CEO blog series
- Daily social media posts
- A 6-part video series featuring tips
- A 12-part email nurture stream
- Weekly podcast
- Weekly webinar
- LinkedIn and Google paid ads
- ABM-centric outreach via LinkedIn InMail
- Hiring manager and candidate guides, e-books, and tips and tricks content

During the campaign period, they continued to arm sales with personalized approaches to prospects. It let sales work as an extension of all the marketing outreach. It included:

- Webinar replay links
- Weekly press roundups
- Relevant case studies
- Social media shout outs for their clients in the news
- Rapid response packages that would get sent to "hot" prospects

The results speak for themselves. They set an all-time record for sales qualified and sales accepted leads in April and May as well as a near all-time record for net new opportunities.

- Website traffic was up 100%.
- SEO was up nearly 100%.
- New client wins in that period: Macy's Stitch Fix, Blue Cross Blue Shield, Target, Cargill.
- Existing clients Walmart, Amazon, and Sea World hired hundreds of thousands of team members using Modern Hire.

Most importantly, revenue growth was double-digit at a time when all their competitors went through layoffs.

Results like these is precisely what every business wants from marketing.

Strategist

It's worth noting some best practices Jay's team exhibited that relate to product marketing when crafting a product go-to-market:

- They took the time to be thoughtful, and then acted swiftly and with confidence.
- This was not about the product, but leveraged product differentiators.
- Everything was customer-centric.
- They chose to emphasize the perception of value when everyone else in the category went to a free trial.

- Their approach was multifaceted—they used everything in the marketing tool kit.
- Sales acted as an extension of marketing and vice versa.

These are great marketing best practices at any time. But if you're finding yourself without a lot on the product front, there is still a tremendous amount of strong marketing possible with a thoughtful approach.

Strategist

Chapter 19

The One-Sheet Product Go-to-Market Canvas

One recent fall morning, the product team at Bandwidth was concerned with solving a thorny problem: the top 10 accounts at the company weren't growing as fast as the next 50. Here's the kicker: the next 50 accounts had vastly different needs than those of the top 10. This was causing friction around how to prioritize development and go-to-market.

John Bell is a smiler and exudes level-headedness. It has served him well leading products at Bandwidth for a decade. As he listened intently to the team discuss solutions, it was clear that for long-term growth, they needed to focus more on companies like those 50—and really push for better alignment between product and all the go-to-market teams.

Bandwidth provides technologies that power almost anything voice- and message-related over a data network. Their products are highly technical, deep in features, span multiple product lines, and have complex, long sales cycles and big competitors. They are precisely the kinds of products that benefit most from strong product marketing.

But at that time, product marketing was a lean team and not consistently empowered to act with clear purpose. They tended to focus more on tactics and tools. It's not surprising (or uncommon) that go-to-market misalignments were more visible to John's product team—which had much deeper category and product expertise—than to those in product marketing.

It's why the product team was the obvious choice to pave the way for a product go-to-market working session to coordinate the go-to-market teams. John brought together leaders from sales, marketing, product marketing, and business development and told them to come ready with market insights about their part of the business.

Three hours later, they emerged with a product go-to-market canvas now owned by product marketing. They felt more aligned, inspired, and confident in their go-to-market strategies and key activities. The product team also gained a more market-oriented framing for what they should prioritize and why.

This chapter introduces you to the one-sheet product go-to-market canvas John used to drive this alignment across product and go-to-market functions. It is where all the levers and concepts we've covered in part 3 show up in a plan. It's the easiest way to create a thoughtful product go-to-market plan and be the strategist so important to product marketing.

Product Go-to-Market Canvas: Think Puzzle

A product go-to-market is strong and strategic when all its activities line up to achieve larger goals that incorporate current market realities. This is why strong PGTMs shift over time.

A product go-to-market is like a puzzle. A product go-to-market canvas is like giving everyone the picture of what that puzzle should look like before anyone starts building it.

And like puzzle building, it goes better and faster with the border pieces in-place first. They create reference points that make filling in the rest much easier.

I created the product go-to-market Canvas (Figure 19.1) to make planning much easier for go-to-market and product teams. Marketing teams get clear on the bigger picture and *why* and *when* for activities, and product teams have more confidence in the go-to-market because of it.

Product Go-to-Market Canvas Example

	Q1	Q2	Q3	Q4
Customer/Outside Environment	• CES in Las Vegas • New computer owners	• Dads & Grads • Gartner Symposium	• Back-to-School • End of Fiscal Year	• End of Year • Holiday Season
Product Milestones	Improved import of existing files	Social features	Features for vertical	Mobile work features
Strategies		**Key Activities**		
Reinvent Productivity: Why you need the cloud for apps	• Carrier deals for pre-install	• End-user success stories • Research on productivity gains results released	• Customer migration stories	• Holiday gift lists on the mobile app • Analyst briefing prep
Teach them to become engaged power users	• Real-time collaboration animations in landing pages	• In-app tutorials highlighting social features • Major social campaign	• Launch new vertical release	• CES event prep for vertical
Embrace and migrate competitive product users	• Gallery of imported document visuals	• Videos on social channels showing old vs. new app experience	• Competitive switching campaign	• Field sales push with special pricing for mobile workforces

Figure 19.1 The Product Go-to-Market Canvas template with a productivity app used as an example.

Strategist

It gives teams the equivalent of puzzle borders that:

- Quickly show opportunities or misalignment—for both product and marketing
- Keep everyone focused, even when improvising or reacting to something new
- See the most important activities at-a-glance and communicate them internally
- Understand the *why* behind actions and stay aligned at the strategic level, not just ticking off tactics

It's intentionally a lightweight canvas, not meant to be a comprehensive plan. Think guardrails that clarify purpose and priorities and keep things on track.

The canvas is a living document, owned and driven by product marketing. It begins with product, marketing, sales, and customer success getting together to discuss:

- The customer's reality
- Competitive and external environment (real life outside of tech)
- Anticipated product milestones/releases/commitments
- The resulting marketing strategies
- Key activities

In the ideal world, the product marketer creates a product go-to-market canvas before the first marketing task gets done. Here's the process.

The Setup: A roughly three-hour meeting. The first hour is about what's working, what's not, market realities for customers, and understanding what must be improved for the business to grow. Each function must share what they've learned for their respective domains. It starts to make clear the gaps the PGTM must address in some way—be it in process or action. The next two hours are about the rest of the framework. This is also where the team can do a SWOT (strengths, weaknesses, opportunities, threats) exercise to ensure you're taking advantage of opportunities and strengths and not defining strategies or tactics that are hindered by weaknesses.

Step 1: Customer and outside environment. Everyone contributes their customer and market knowledge. The point of doing this first is it grounds everything from the perspective of customers, not the company's. Add anything that might influence activities: Competitive announcements, conference events, major ecosystem releases (like iOS updates). You don't need to list everything—just key elements that might affect what people think, do, or see. This goes in the top *Customer/ Outside Environment* swim lane.

Step 2: List any known product milestones. With agile, you may not have a lot of detail, but for major initiatives or commitments—integrations, new platforms, features for particular customer segments—you probably have some direction. Capture whatever is known in the *Product Milestones* swim lane.

Step 3: List your marketing strategies. In this canvas, you assume business goals are those of the company. Strategies are the guardrails for activity. You should be able to draw a direct line between your strategies and how they move the needle on business goals. Clear strategies help those doing the actual execution understand the bigger picture their work supports.

Step 4: List key activities supporting strategies. This is not everything on a team's checklist—it's only important activities others may be dependent on or want to plan around. It might include a quarterly sales training, defining a new sales playbook, launching new pricing, naming key product features, or building a demo across a suite of products. This canvas is as much a communication tool as it is a planning one.

Strategist

Be Sure to Leverage Your Strengths!

An important part of Step 4 is understanding your strengths and weaknesses for strategies to be appropriate. I'm a fan of SWOT (Strengths, Weaknesses, Opportunities, Threats) because it's a simple, fast gut-check on whether you're defining strategies that leverage your opportunities, avoid marketing right into the teeth of a competitor, and leverage your strengths while not requiring you to fix all your weaknesses.

For example, if a weakness is your company has a very small sales force relative to the competition, you may decide to prioritize enabling partnerships or to do more product-driven growth.

Step 5: Work from the outside edges in. Start with how you want the year to end and work backwards. Do you want to have more partners by the end of the year? What will they need to see when, and what's a necessary precondition for anything you want to do? This helps make sure the right foundation is in place for what you want to build.

The tendency of most is to start with the most immediate quarter or two, fill in all the things they think they need to do, then the canvas peters out in the outbound quarters. This is especially true for early-stage startups; there isn't much value planning past two quarters. Things change way too quickly to plan beyond that.

But it's still important to start with the end in mind to make sure you don't miss something that's not already on your list. It might also make things fall off your list. For earlier stage companies, just bucketing "second half" as a parking lot for ideas works fine, but don't skip discussing it.

As you're filling in activities, look at the customer and outside environment to intersect key activities with their realities as well as what's happening in product.

Step 6: Revise. In your first meeting, you'll get solid direction of where things need to go, and most importantly, alignment across functions. Use time outside the meeting to think through details and fill in gaps. Don't try to fill everything in. You want to have some capacity to add things you haven't anticipated and keep the canvas responsive. I recommend bringing the teams together to revise thinking once a quarter.

Make It Customer First!

You can do all the "right" things but still not achieve your goals despite the effort. What often separates a really good product go-to-market canvas from others is how well you're mapping actions to your customer's world.

Everyone gets how critical this is for messaging (we'll spend all of part 4 on how to do this better). But it's equally important for how PGTM actions are planned.

Table 19.1 shows the exact same scenarios with a company-first lens and contrasts it to a customer-first lens so you can see the difference.

Table 19.1 PGTM canvas work helps you stay customer-first, but know the difference.

Scenario	Company-first	Customer-first
Major new product release is ready in March.	Launch when product is done, which turns out to be Spring Break for many.	Work with product to find a launch date that is clear of most major holiday breaks and in time for an industry conference.
New customer vertical campaign goes live in December.	It took two extra months to get the agency to complete the work and for internal teams to be ready. Sales worries they won't make their number without the new vertical, so the campaign launches in December.	The new vertical is busy with their own end-of-year planning and the holidays. They don't have time for something new. Focus instead on a year-end campaign with a pricing incentive. Launch the new campaign with the new year when the vertical has more capacity for it.

(continued)

Anchoring PGTM from the customer's vantage point is about optimizing based on what customers want to hear and when they have the capacity to hear it. Sometimes who does the talking matters as much as what's being said. If a sales rep tells a customer about a coming product upgrade it lands differently than if a customer success manager says the same thing. Thinking this through is crucial in making anything in your PGTM effective.

PGTM Canvas in Action

Let's dive into the details of that initial product go-to-market kickoff with all of Bandwidth's teams. In the first hour they determined:

- They didn't have a strong enough brand or clear market position.
- Their fastest growth customer segment was one that sales knew little about. The company needed to do a better job organizing around their target market segments, not just their products.
- None of their externally facing digital assets did anything to speak to their high growth customer segments or how widely varied the customer journey was for the different product lines.
- Sales front-line experiences weren't getting back to product or product marketing. The latter, in particular, wasn't co-creating key content with sales.
- They weren't learning from customers they lost.
- They weren't clear on the ongoing value they could provide customers after the initial sale.

Figure 19.2 shows how the framework started to take shape to address these gaps and the business goals.

This is as far as the initial meeting got and was all that was required. The first meeting is about capturing ideas and needs. The customer-first polishing and idea refinement happens in smaller, subsequent meetings.

Strategist

	Q1	Q2	Q3	Q4
Customer/Outside Environment	• New regulations • March Madness	• Important industry conference • Gartner MQ for major product	• Gartner Symposium • Zoomtopia • MSFT Ignite	• Dreamforce • AWS re:Invent • Election • Holidays
Product Milestones	New feature for regulations	Integrations with new Cloud comm providers		Better performance across products
Deepen engagement & share of wallet and differentiate our brand's feeling as a trusted partner	• Convene customer advisory board to enrich feedback and deepen segmentation	• Guest post from Cloud comm providers' engineers • Star Developer circle	• Customer appreciation and delight campaign	• Annual planning review w/CS team—where can you adjust services?
Become a thought leader on the regulations and protection of user experience	• Educate space on new regs and product implications for prospects • Micro-video "new laws" from 3rd parties	• Educational series on new regulations	• Customer communities/forums for shared learning	• Protect your customers from election spam
Become the leading provider powering Cloud communication systems to the most innovative, fastest growing companies	• Deep customer research and segmentation work • AR effort from PMM/PM teams • Deepen customer stories	• Update all digital assets to reflect audience • Analyst trends breakfast	• Showcase at Ignite • Product proof showcase of service improvement	• Special EOY pricing to upgrade into more of product line

Figure 19.2 The resulting PGTM working session with go-to-market teams.

157

Putting everything into a product go-to-market canvas was eye-opening for every go-to-market function. They saw how they could shift people, process, and tools to better serve the business. The product team also felt more confidence that go-to-market could support the business and leverage the product decisions they were making.

Most importantly, it created alignment across functions when there wasn't enough before. That's the primary job of the product go-to-market canvas, which then sets a better frame for go-to-market planning.

One thing it's not is a complete marketing plan. I explore the difference by walking through examples of them next.

Chapter 20

Understanding in Action

Real Marketing Plans

A product go-to-market canvas is a tool that drives alignment between product and marketing activity. It sets the product-oriented, more strategic frame for the detailed plans that marketing teams own and execute.

Product marketing work gets embedded in larger *marketing plans* that cover *all* marketing activities. But if your day job isn't looking at marketing plans, it's hard to know if the strategic thinking in a product go-to-market canvas is being carried through in how the marketing team planned its activities.

This chapter looks at marketing plans from real companies, so you can see how product go-to-market work is reflected in marketing plans. You'll notice (not surprisingly) that the plans get better as company stage matures.

Early Stage

For earlier-stage companies, the product go-to-market canvas and marketing plan should be really close to one another. When they diverge, it can be a signal that marketing isn't keeping all activities aligned. That said it's also a very dynanic time. It is equally important to try very different marketing activities so you learn which work best for your company.

This company had one product and additional add-ons. It had around 20 customers and more than $1M in annual recurring revenue when this marketing plan was created.

Goal: Transition to early majority through brand and marketing

Key Result: 40% Marketing Qualified Lead to Sales Qualified Lead increase

Objectives:

- 60% of customers willing to be in referral program
- 70% increase in inbound leads
- 1 closed deal > $100,000 from major event
- 3 customer webinars, speaking at 12 events, 3 placements of thought leadership from customers

This was part of an 80 slide presentation (not including the appendix) that justified every aspect of what was being recommended. First, that's just way too long. It's easy to think the clear, measurable objectives get the job done. They aren't enough. Let's examine ways this plan could be better.

Measurable Goal That Ties to the Business

The most important goal at this stage is establishing a customer base that enables the future business. That could be customer numbers or revenue numbers, but either way, there should be a specific target that aligns with what's been identified in product go-to-market work.

The plan also doesn't have an objective about the product holding a market position that helps it win. Investing in this foundation is crucial even at the early stages.

That aside, remember our technology adoption curve? There isn't any objective tied to the right kinds of customers becoming leads. As it relates to the defined goal, if this company is early stage, they aren't anywhere close to early majority.

Key Results That Pair Quantity with Quality

The strategies and tactics need to include activities that refine the target customer segments, not just focus on getting in leads. These are all areas where product marketing works tightly with marketing in iterating the plan.

Even if marketing is hitting its Marketing Qualified Leads (MQL) to Sales Qualified Lead (SQL) goals, it doesn't mean the business is healthy. You want key results to also indicate quality of pipeline. This is as important as quantity. For example, goals around sales cycle time, average contract value, and the win/loss rate can all be quality indicators of what's flowing into the pipeline. Product marketing work affects all these metrics.

Objectives, Strategies, and Tactics

The objectives listed are all actually key results. Knowing the difference between an objective versus strategy or tactic can be confusing because it depends on stage and business context.

For example, "Become the preferred platform ecosystem" for a company at scale is a strategy. It might include tactics such as:

- Establishing or enhancing a partner program
- Engendering loyalty among the most productive partners

The end-of-year key results for that strategy might be:

- Five new tier-1 partners added
- Overall partner network grown by 25%
- 100% increase in data processed via partner APIs

But if you're just starting out, the strategy might be what was a key result for a company at scale: "Get five new tier-1 companies as integration partners." Some partners carry so much influence in a category that landing them is a strategy unto itself. Tactics wouldn't focus on creating a general partner program that works for many. Instead, they would be much more specific like:

- Promote an improved integration API specifically for that short list of target partners.
- Create an engineering blog series on how those integrations were tested and proven to improve performance.
- Get five "friendly" partners who can act fast. Have those engineering teams post their impressions and questions on developer forums.

When a company is just getting started, every aspect of figuring out go-to-market is massively fluid. Good early-stage marketing plans set the strategic frame and identify key important actions but leave a lot of space for what might happen. They are also clear on what to measure so the company knows if it's succeeding in marketing, and the measure of that is if the business is succeeding.

Bottom line is this was way too much specificity on the wrong things for this stage.

Scaling Stage

This company had more than $10M in recurring revenue. They were trying to refocus their efforts to where they could win and improve their growth rate.

Goals
- Define the category and establish our company as category leader.
- Create a demand creation operation to predictably deliver qualified opportunities to sales.
- Harness the partner ecosystem to achieve category and demand creation goals.

Strategies by Quarter

- Build foundation (Q1)
- Establish baseline/start operationalizing (Q2)
- Marketing mix solid/continue refining high potential channels (Q3)
- Monitor and optimize (Q4)

Here is what could have made their plan stronger:

Tie Goals to the Business

A goal tying marketing's plan to the business was completely missing. *Drive pipeline and improve conversion to meet revenue goal of $20M.* At the growth stage, revenue is one of the most important indicators of a company's success.

Strategies versus Tactics

Strategies should never be your to-do list of what should get done by when. Likewise, it shouldn't be a strategy to monitor, optimize, and adjust; that's table stakes for doing the job.

Better strategies would be:

- Elevate industry and customer validation to enable evangelism.
- Optimize key stages of funnel and sales process.
- Increase experimentation across channels to discover new ways to find target customers.

At this stage, there is usually some awareness of your company in the market but likely the picture of your product and what it does remains muddy. There might be lots of competitors saying similar things and no one has yet pulled ahead, or you want to solidify your lead in the marketplace.

Look for what best amplifies your market position. This is where product marketing work should be a major factor. Focus on customers that best represent how you want to grow your

Strategist

business. It might be time to be much stricter in qualifying customers that make it through your sales process. Not all customers are equal!

It's important to figure out which are the best customers for your business. Make finding and converting them repeatable during this stage. The product marketing work that contributes to this focuses on enabling sales, evangelism, and exploring new distribution channels.

Mature Stage

This tech company had hundreds of millions in revenue and a dozen products in the market. Their highest growth businesses were in newer product lines being integrated into the company's flagship products.

> **Goal:** Grow new product line revenue as percentage of overall revenue by 20%.

Strategies
- Drive awareness and adoption of product line solution.
- Establish company as leader in new categories.
- Drive awareness of how company enables service delivery for target audience.
- Expand meaning of company brand to target audiences.

This company was already very successful. You might think this is the time when a marketing goal separates from the company's.

But in this example, the marketing team still used the company's goal to drive the marketing plan. This is particularly useful at tech companies when sprawling marketing organizations with huge budgets get a critical eye from other functions. Goal alignment shows why marketing does what they do and how it serves the business.

Strategist

The actual plan had specifics I can't include: they named the category they wanted to evolve, they articulated which target audience they wanted to get better at being relevant for, and they had a name for their product line that also helped sell it. All those specifics made it good and heavily leveraged strong product marketing work.

An equal part of the work for an organization at this scale is how to roll out and evangelize support of this plan among the other groups essential for the plan's success. Likewise, objectives and key results, whenever relevant, need to be aligned across functions so each group's success is linked to those they depend on.

Pro Tips in Crafting Good Marketing Plans

These tips apply across any company at any stage.

Define the playing field. I nearly always see a "lead or define category" variation in marketing plans. Most know how to declare and build a case for their point of view, but don't skip defining the playing field. What should or shouldn't be considered? Help people evaluate how to think about a solution or a space. Know that defining the playing field and your position in it takes time. Also, keep looking at what your competition has done in the space. Plenty of companies failed because they didn't realize the goalposts of success were moved by another company.

Beware the allure of paid marketing and big budgets. Access to resources can be both a blessing and a curse. Paid marketing can mask if the organic foundation for growth isn't really in place. Without healthy organic growth, you're on an endless spending spree where growth only comes with more money or resources. There are many ways to build this foundation: content, social, comparison sites, pundits co-promoting a webinar, participating in digital forums, making customer testimonials more visible (and less produced).

Strategist

When it comes to marketing plans, the best intersect all the strategic go-to-market thinking already done by product marketing. It helps marketing track to the way *product teams* envision the business growing given what they're building. Detailed marketing plans then bring all the go-to-market magic to life through the GTM engine.

Part Four

Storyteller:
Clarity and Authenticity:
The Process and Tools to Rethink
Messaging

Chapter 21

Discover Your Position

At a cocktail party one evening at RSA—a massive annual security conference—a prominent cybersecurity leader confidently worked the crowd. A "made man" in security, senior executives from a wide range of companies walked up to him with warm hellos and genuine interest, asking, "What are you up to these days?"

The executive, Brendan O'Conner, was the former Chief Trust Officer at Salesforce and Security Chief Technology Officer at ServiceNow—titans in their respective industries. After seeing first-hand what was lacking in cloud security, he decided to create the solution and co-found a startup, AppOmni.

As often happens with startups with a new way to solve a problem, they wanted to describe what they do using equally new language. Brendan wasn't yet sure what would work, so he went to the party testing a prepared talk track on each person who asked him what he was up to.

Nearly everyone's response to any variation he tried was, "Is it similar to what [company X] is doing?" or "Oh, you mean like what [existing cloud security tool] does?"

When it came time to process the evening's learnings, it was clear that even for people with deep knowledge, talking to a person they respected and trusted, they made sense of AppOmni using what they already knew. It was a powerful reminder that people need to start with what's familiar to understand what's new.

Brendan was a bit surprised by how much he had to position his brand new product using existing products as reference points. But he continued to listen, learn, and adapt AppOmni's messaging based on every conversation he had in those early days.

Iterative learning like this lets you discover what customers need to hear. You're looking for gaps in market understanding so you know what your messaging must bridge.

Chapter 2 shares the narrative we crafted for Word, making it the best reviewed word processor of its time. It framed a counterintuitive approach (fewer features) as better by using key messages (focused on how people actually use word processors) to support our story.

Positioning and messaging work best when part of a bigger story. It gives people more reasons to believe. Stories are also more memorable. It frames the product's position by making messaging feel like it is information with purpose. No one feels sold; they feel more knowledgeable.

A brief recap on the often-confused differences between positioning and messaging:

- **Positioning** is the place your product holds in the minds of customers. It's how customers know what you do and how you differ from what's already out there.
- **Messaging** includes the key things you say to reinforce your positioning, making you credible so people want to learn more.

Positioning is your long game. Messaging is your short game. The later adjusts to what is relevant to customers at a given moment, for a given context, or for a specific campaign. In aggregate, it reinforces the desired position, but its use can be more specific to its purpose.

Positioning Takes Time

When Microsoft Office was first introduced to the world, we changed every reference of *desktop productivity application* to *integrated office suite*. It was part of a systematic effort to reinforce the story we wanted told: integrated office suites were the category.

I was sick of writing and saying it after a couple years and remember thinking *surely the market knows this by now?* But our sales numbers told another story. Our biggest revenue months happened just as we were about to launch our new version nearly two years later.

Think back to the technology adoption curve—the dogged positioning investment we made set us up to accelerate more quickly through the fat parts of the bell curve. But it took years.

Positioning always starts with elements of messaging, but it's made real through the combination of all the activities reinforcing a bigger story. For example, if you have a direct salesforce, a lot of positioning happens in the back-and-forth of an evaluation process. An assessment guide with strict evaluation criteria—reinforcing your position—frames how customers look not just at your product but at the competitors. Likewise, how you train sales to respond to a question about the competition can create positive (or negative) bias for the rest of the buying process.

Positioning also happens organically through reputation. All the places where evangelism lives—or is noticeably absent—influence market perception. Word-of-mouth elements like comparison sites, reviews, ratings, social postings, shares, online forums, other people's content, and even employee buzz all play a role. That's just what's seen. There is also the unseen activity that can carry even greater influence.

This collective sentiment has enormous impact on how the world thinks about a product.

For all these reasons, positioning should be thought of as the outcome of all the activities done for a product

Storyteller

go-to-market, not just its messaging. But because messaging is within your control and so essential to telling the story of your product, it's the focus for the rest of this section.

Good Messaging Is Harder Than It Looks

Messaging is different from what most people think. It is not a catchy tagline or pithy statement of benefits nor is it a statement of a product's positioning. It's not birthed in a room by marketing.

Bad product messaging is easy to spot. After reading it, you still don't know what something does.

Good messaging feels natural and obvious. It's harder to see what makes it good: anticipating what people need to hear—be it *just the facts* to something more aspirational. It also places a "You Are Here" sign in people's mental maps—as Brendan discovered, a really important part of the work.

Done well, good messaging is the outcome of a market-driven process in which a product's position is clear. It frames a product's value and makes people want to lean in and learn more.

Formulating it is a process, starting with Fundamental 1: deep insight into customers and the market, followed by a lot of discovery work, which I talked about in chapter 11. You're trying to understand how prospective customers see *their* world. What do they *already believe* to be true? Discover *their* knowledge gaps. You want your messaging to bridge them.

It's important to remember people are skeptical of anything you say. They have every reason to be. New technology always comes with a cost beyond a product's price. People and time aside, it's yet something else to manage. And far too many times, products don't live up to their promises.

Part 4 of this book focuses on how to create messaging that is more meaningful. I'll go deep on examples that eschew traditional formulas and show how they evolved over time to reflect changing market and business dynamics. I'll explore how to balance what you learn in your discovery processes with final messaging and get the entire go-to-market engine using it with a one-sheet canvas.

Accuracy for the Engineering Trained

For people who come from engineering backgrounds, the need for precision is real. It's what makes statements feel believable. Anything less can sometimes feel less honest or incomplete.

Know that talking about something in a way that simplifies—and leaves some details out—is not being inaccurate. It's guiding toward shared understanding. When gravity is introduced as a concept, most start with Newton's story of an apple falling from a tree before introducing:

$$F = G \frac{m_1 m_2}{r^2}$$

Messaging's job is to connect first before diving into all the details. It's hard to process depth without some table-setting context. If everything said about a product is true, it is still accurate. It might not be the precise way you would talk about it with someone with deep knowledge and full context, but that is the precise difference between the job of messaging versus that of product details.

Storyteller

Chapter 22

How to Listen and Connect

Expensify and Concur

In my product marketing workshops, I have an exercise where I ask the group to choose from a list of messaging examples and select the one that best connects with them as well as the one which makes them curious to learn more.

For over a decade, the overwhelming winner on both dimensions was this one from the early days of Expensify:

> *Expense reports that don't suck! Hassle-free expense reporting built for employees and loved by admins.*

We always discuss what people love about it. It's some mix of:

- **It feels emotional and true.** Expense reports do suck! When they read that this product is about making them not suck, it feels believable because they were willing to say how we *feel*, not just talk about what the product does.
- **It efficiently communicates benefits in simple and compelling ways.** "Hassle-free" doesn't specifically reference automation or improved efficiency, but it immediately conveys ease for the people that do—and dread—expense reports.
- **It covers multiple interested audiences.** It conveys a sophisticated understanding of expense reports—that

174

they are also a lot of work for admins—financial and administrative—not just the people wanting reimbursement for themselves. Their product works well for all interested parties, and they manage to convey this in just four words "and loved by admins."

Every word in their messaging means something concrete and packs a thoughtful *and* emotional punch. It was so successful in its day, it had the ultimate compliment: lots of copycats, like Zoom that went with "Video conferencing that doesn't suck."

We can also infer from this messaging the approximate market position they wanted to own: the easiest, most loved way to simplify expense reporting for everybody who touches it. But notice how they didn't use this as their messaging.

In the exercise, I tuck in messaging from attendees' competitors. Most look like this one straight off a home page.

> *[Our product] improves the efficiency and effectiveness of field service organizations and mobile workforces. Meaningful, measurable value that optimizes your business, improves efficiency, and delights customers.*

How many products could you substitute into this statement? Hundreds.

Do you have any ideas on what position they're trying to own? I don't.

Unfortunately, most messaging is closer to this last example than to Expensify's. Good messaging doesn't just say what a product does or its hoped-for benefits. It conveys a deeper understanding of the people the product is built for, which requires great listening skills. You're looking for insights that make customers feel known.

The rest of this chapter shows you how.

Storyteller

Listen and Learn

Everyone wants to feel known and seen. But most of the time, we are bombarded with what people want to say, not anything that reflects they know us.

Just like Brendan's colleagues at RSA, everyone is trying to put a product into their situational context. Good messaging anticipates this and provides some of that context as a connection point.

Getting there takes an unvarnished understanding of your target market's daily lives. The best way to get this is through direct conversations with them. Ask open-ended questions like:

- Tell me what your average day looks like.
- What really frustrates you in how you do your job? What keeps you up at night?
- What is the straw that broke the camel's back? What made you seek a new way of solving the problem?
- If you had a magic wand to create something that could do anything, what would it do?
- What's the last problem at work you spent money to fix?

Listen for insights and language that makes them feel known, like you really get how they're experiencing a problem. You can do this as part of a product market discovery process or as part of regular customer discussions with either prospective or recent customers. This is one of the reasons why I recommend weekly contact with customers.

It's only after understanding customers' deeper context that you discover insights that lead to truly compelling messaging. The next step is to test different messaging directions based on what you've learned. What resonates might surprise you.

Storyteller

Choose Credibility and Clarity

Expensify still competes with seasoned incumbent Concur, which not only had a 15-year head start on the then much newer Expensify but was acquired by SAP. Expensify still occupies the position of being the more modern, innovative, and faster-moving disrupter.

In Table 22.1, I'm putting more recent messaging side by side so you can see how they say the same thing but use very different language and style. We can infer from both that those who feel the pain the most travel a lot and want to manage expenses.

"Best messaging" isn't an absolute. You want messaging that both performs well for your product's go-to-market strategy *and* positions, which means choosing one that balances those needs accordingly.

SAP might be focused on creating alignment across their enterprise divisions, and driving consistency was an important factor in what was approved. Let's also assume their messaging was tested and the versions we see here converted better than others.

Table 22.1 A tale of two messaging styles.

Expensify	SAP Concur
Sit back and let Expensify do the work.	Better travel and expense management. Always.
Whether you're a road warrior with pockets full of receipts or a busy accountant buried in paperwork, Expensify automates the entire receipt and expense management process. • One-click receipt scanning • Next-day reimbursement • Automatic approval workflows • Automatic accounting sync	Helping businesses with their top travel and expense challenges. • Concur Expense. Submit and approve expenses from anywhere. • Concur Travel. Capture travel no matter where it is booked. • Concur Invoice. Automate and integrate your AP process.

But just as with the building of products—how data informs but shouldn't drive the outcome—clicks and corporate directives should not be the only determinant of a product's messaging. The unintended consequence of creating messaging that way is a product's position and differentiation can become less clear over time. Messaging should deliberately serve go-to-market strategy. It matters.

Acknowledging that there may be such unseen influences in both cases, let's dive into each further.

X-Ray: Concur

Better travel and expense management. Always. → They conclude they're better but don't say why. They also treat the employee and admin as experiencing the same benefits. On the plus side, what they do is clear.

Helping businesses with their top travel and expense challenges. → Being helpful is about as run-of-the-mill as benefits get.

- *Concur Expense. Submit and approve expenses from anywhere.*
- *Concur Travel. Capture travel no matter where it is booked.*
- *Concur Invoice. Automate and integrate your AP process.*

The list is led by individual product names, not the customer's point-of-view. They definitely are trying to emphasize mobile, cloud, and integration through "from anywhere" and "no matter where it is booked."

We're left being asked to believe something is "better... always" because it can be done from anywhere and integrated into existing processes. That's if we get past the list of products to really try to understand what they mean.

X-Ray: Expensify

Sit back and let Expensify do the work. → This sets up the idea that the product is going to do our work for us. It infers ease and automation but doesn't say so as the lead.

Whether you're a road warrior with pockets full of receipts or a busy accountant buried in paperwork, Expensify automates the entire receipt and expense management process. → The first part really lands because of specific details that capture real peoples' frustration. When I'm traveling for business, I have receipts coming out of my wallet, backpack, and suitcase. Our CFO always sends out a "last call!" email to synchronize an avalanche of digital paperwork with the accounting system to close that months' books.

It *shows* they get peoples' reality. Then they tell us they automate the entire process before diving into how they do it:

- *One-click receipt scanning*
- *Next-day reimbursement*
- *Automatic approval workflows*
- *Automatic accounting sync*

This list tells us what Expensify does that delivers against its promise with specifics—"one-click" and "next-day" are notably better than the five-clicks and 30-day norm many older systems require.

Small but not insignificant: language like "accounting sync" versus "automate and integrate your AP process." They mean the same thing, but the former is more efficient and modern, the other is precisely correct but feels tired.

The workforce is now more than 50% millennial with Gen Z right behind. *Messaging needs to match the temperament of the generation it's trying to reach.* Whenever possible use evocative, energizing language that also helps you connect with your target market.

And the only way you'll know? By listening to them talk about their lives, then testing messaging so it achieves your desired results.

Storyteller

CAST: A Simple Guide

It's easier to see good versus average messaging in someone else's work, but much harder to recognize it when you're creating your own. That's why I introduced CAST briefly in chapter 6 as a gut-check as you develop your own messaging. Grade yourself on each of these guidelines and push for better.

1. **C**lear. Is what you do clear and is there a reason to be curious? Is being comprehensive getting in the way of clarity?
2. **A**uthentic. Is the language evocative and meaningful to your customer? Is it said in a way makes them feel known?
3. **S**imple. Is it easy to understand what's compelling or different? Will customers know what's better?
4. **T**ested. Was it tested and iterated *in the context customers will experience it?*

It's worth mentioning again how important testing in the context in which a customer will experience it is. It immediately makes flabby language easier to spot. You also get a sense of what you can use in design or visuals to emphasize points, so not everything has to go into product messaging.

If you review each Expensify example, they do well by all these measures.

The best companies—and certainly any new challenger aiming for disruption—find ways to tell their story through messaging that is clear and grounded in what customers already believe. It positions. It inspires. It's the foundation for a product's story. It's why I dive so deeply into messaging as part of product marketing Fundamental 3.

Chapter 23

Understanding in Action

Netflix and Zendesk

My kids can't comprehend that I watched *Friends* over a decade of committed Thursday night viewings. They got to know that Ross winds up with Rachel in just a few months. Their entire lives have been lived in an era where TV is watched on demand, on any device, and with total customer control. The biggest force behind this massive transformation in consumer behavior was Netflix.

Zendesk started 10 years after Netflix. Although SaaS had been around, the idea of going directly to customer service reps with a free trial—as easy as Netflix's—to sell a product intended for big businesses was relatively new. Zendesk decided there were plenty of companies who wanted simpler interfaces and a more modern way to buy. They became one of a wave of SaaS companies transforming how B2B software goes to market.

Both companies had exceptional products and strategies *combined* with exceptional messaging. It positioned their products well for the market and their business.

Their go-to-market models depended on converting visitors into trial customers. They messaged accordingly and *evolved over time*, adapting to shifts in customer behavior, company awareness, and business strategy.

This chapter looks at how all these shifts were reflected in their messaging. I'll explain how they meet CAST guidelines, but take as a given that everything on a page was heavily tested and iterated. Both companies famously treated their websites as products.

Netflix's DVD Days

Before movie and show streaming, the standard way someone watched a movie or TV series was going to their local video store to rent VHS tapes or DVDs. The problems?

- *Cost.* $4.99 got you *one* movie, and with late fees it didn't take long for one movie rental to cost a lot more.
- *Inertia.* To get a movie, you had to drive somewhere, pick something out, and wait in line. Talk about inconvenient.
- *Choice.* Most rental chains did not carry a lot of classic movies. They were also always out of new releases. And the DVD of the TV episodes you wanted often wasn't there when you needed it.

Netflix didn't start by messaging a mission to transform home entertainment, even though that was their long-term vision. For more than their first decade, their messaging focused on exactly what their service did and why it was better than the status quo (Figure 23.1):

- "Rent as many movies as you want! For only $8.99 a month." They made *clear* that for the price of a couple rentals, you got as many as you want.
- "Classics to New Releases to TV episodes." Embedding the problem in how they talked about their service ***authentically*** captured what was wrong with the status quo. It showed why they were better in a ***simple*** way. They could have said *thousands of movies and shows to choose from* but didn't.
- How it worked was more convenient: "Free DVD shipping—both ways." Your movies getting shipped to

Figure 23.1   Netflix's website in 2009.

you made it *simple* to understand how this was different from driving to rent a movie.

- Their third supporting bullet was not a benefit or about their product. It was about mitigating the perception of risk: "Cancel anytime." For a newish Internet company, it was important to build trust. That's also why a 24/7 customer support number you could call was highly visible on the page. The transparency was an *authentic* way to build trust.
- The visual surrounding the messaging was the promise of snuggling up at home with the family for a fun movie night. It did as much as the words and packed an emotional punch.

As the Game Changed

Netflix streaming was an enormous strategy bet that transformed their business. It was important to move consumers to streaming quickly. Even though this shift would take time,

Storyteller

Figure 23.2   Netflix's website circa 2014.

their messaging firmly moved toward what was important to the business (Figure 23.2).

- "Watch TV shows & movies ... " TV now leads because it's what drives "bingier" behavior, bringing people back again and again. They are *clear* on what you get and make customers feel known (*authentic*) by acknowledging shifts in viewing behavior.
- "...anytime, anywhere" messages in a *clear* and *simple* way the distinct advantage of streaming over DVDs—you could watch on your smartphone, laptop, or tablet in addition to your TV.
- "Only $7.99 a month." Price was a core part of how they conveyed more value than the DVD service, so they made the price *clear* up front.
- "Start Your Free Month" was an invitation to give the service a try in a call-to-action button that made *clear* what to do if you were curious.
- The visual around the messaging changed a lot too. Large, flat screen TVs were the new norm, so the page design reflected this shift. It also let them show the product experience on the TV visual.

Brand Leads the Message

By 2016, Netflix was on a mission to deliver the most talked about content in the world. Beyond Netflix originals, they focused on being the platform where shows like *Breaking Bad* and *Schitt's Creek* became part of the cultural zeitgeist. It's human to want to feel in the know and belong, and they leveraged that (Figure 23.3).

- With their brand and what they do well established, Netflix messaged, "*See what's next,*" a ***clear*** reason to be curious that also ***simply*** set them apart from their growing competition. Netflix messaged an invitation to become part of the cultural zeitgeist.

Figure 23.3   Netflix's website around 2016.

Storyteller

- "Watch Anywhere." Again a simple promise on what could be done that their product was the best at—seamless watching between platforms and watching shows offline.
- "Cancel Anytime." An *authentic* way to continue to build trust. Likewise, "Join" is used versus "Start Your Free Trial" because it's now more than just a service. You belong to a tribe.

Each of these word choices might seem small, but in aggregate, they pack enormous punch. It positioned what was different and their overall value in remarkably few words.

It took the better part of two decades to get to the place where their messaging could coast off their brand. It's the technology adoption curve at work. Even with the greats, greatness takes time.

Zendesk Anticipates What Customers Want to Know

Customer support software had been around for a long time when Zendesk started. But it was trying to do something very new for its category—grow through word-of-mouth and free trial and not rely as heavily on a direct sales force to grow. That meant its messaging had to anticipate what people wanted since they weren't necessarily going to talk to a salesperson.

You can see this in Zendesk's early messaging (Figure 23.4), where they even ask the question for you.

- "What is Zendesk? *Web-based* help desk software with an *elegant* support ticket system and a *self-service* customer support platform" (underline mine, indicating *simple* differentiators). It also makes very *clear* what they do. What's interesting is this does not lead in the header. Rather, what best positions them is stated *simply*: "*Customer Support Made Easy.*"

Explaining what they did was still necessary three years after starting and 10,000 customers in. This is how long it took the

Figure 23.4   Zendesk's website, circa 2010–2011.

industry to know what Zendesk did. It's yet another reminder of just how long it takes even successful businesses to move through the technology adoption curve.

- Notice the absence of words like *modern* or *revolutionary*, even though they were all these things. Instead, they kept language more **authentic** to what their audience wanted to hear. That made their differences clear.
- They did a lot with the rest of the homepage to feel **authentic** and credible to their target customers. With only a glance, you see customer validation ("10,000+ companies use Zendesk"), it's global (Customer Support activity map), it's multi-platform (product visuals in the main hero image), and that it is easy (no credit card needed). Messaging can be embedded throughout a page and can take many different forms.

Storyteller

They were unafraid to have their tone and brand feel different[1] in their early days and express some love—literally (note the heart in the logo).

Customer Outcomes Lead

With a pending initial public offering (IPO), the category knew what Zendesk did, so their messaging shifted to customers' desired outcome (Figure 23.5).

- "Customer satisfaction has never been easier" gives a *clear* reason to be curious.
- Product positioning got even *simpler*: "Beautifully simple customer service software."
- The rest of the home page evolved and *authentically* engaged with what they knew prospects needed. 30,000+ customers and great brands trusted them. They made calculating ROI easy versus just claiming they had good ROI.

Zendesk made sure to position their competitive advantages well: strong product design and how easy it was for customers to try the product. After their IPO, they doubled down on what made them different (Figure 23.6).

- *Beautifully simple . . . software for better customer service.* They reduced their messaging to what most simply and clearly communicated what they did and how it was different.
- Putting pricing up front without needing to click was an *authentic* way to give people what they wanted to know up front and reduce friction to a trial.

---

[1] They heavily researched their Buddha branding choice, wanting to make sure it didn't unintentionally offend while still standing out. Despite its memorability, it was eventually deemed limiting. They ultimately moved to something more reflective of how their product line evolved.

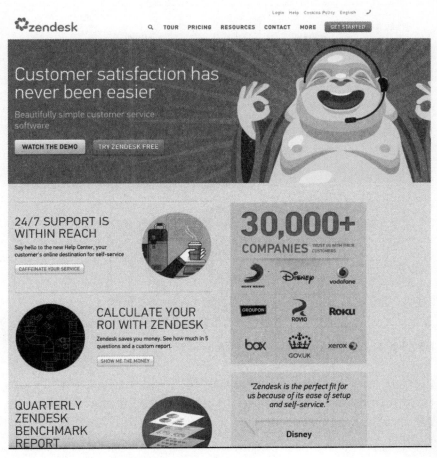

Figure 23.5   Zendesk's early 2014 pre-IPO web page.

In this era, if you compared Zendesk's website analytics—average page views, time on site per visitor, etc.—to any other major B2B SaaS company, they outperformed them all. They took the website's role in the customer journey very seriously, and it was why they were so successful in introducing a new GTM model to their software category.

Both Netflix and Zendesk are best-in-class examples of how great messaging is a combination of what customers need to hear with a company's business strategy and the current

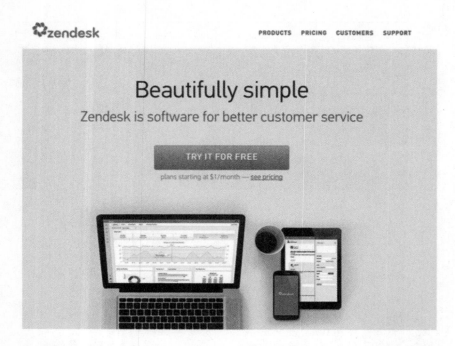

Figure 23.6   Zendesk's website in the initial post-IPO era.

market conditions. Their messaging brought to life what was most important and did so in a way that made it all more meaningful. They made sure it was the right message, for the right time.

Profile in Product Marketing: Julie Choi

Product Marketing for Developers

Back in 2008, iOS was just about to be announced and Android wasn't even born. Facebook had just launched their social app platform for third-party developers and Yahoo was excited to get in on all this action. It's why they hired Julie as their first product marketer dedicated to developers.

In those early years, she learned it was important to approach the audience with humility and a desire to understand their problems. She worked with Yahoo Developer Network engineers to translate the value of APIs, services, and frameworks in a way that mapped to developers' day-to-day problems. She realized the key to connecting was to prioritize developers' problems and offer Yahoo's solutions.

Product marketing to developers became Julie's thing. For the past decade at Mozilla, Hewlett Packard Enterprise, and Intel, she watched the pace of innovation and development only accelerate. Developers enabled the rapid adoption of Internet, Mobile, Social, and Machine Learning powered experiences, which meant they, in turn, had to adapt and retool faster than almost any other profession.

Doing outreach in a world moving this fast means marketing with the highest quality documentation, demos, code samples, and environments for hands-on training. There are more tools than anyone can count, so cut to the chase by showing the code. This shows (vs. tells) how tools make development easier, faster, or less expensive.

In her many years working with engineers, designers, and architects across the world, the consistent theme in developer marketing has been to keep the messaging mix 20% aspirational and 80% pragmatic. Messages with examples keep both sides of the product/user equation honest. This zero-fluff, straight-shooting, bi-directional approach toward product marketing has led to meaningful growth everywhere Julie's worked.

Storyteller

Chapter 24

The Balancing Act

Right Message, Right Time

Back when the Internet was new, Microsoft Office had a handful of features that were Internet-ready. It was a standard pre-launch practice at that time to go around the country doing focus groups to test feature names. One of the exercises asked people to take cards that described what a feature did and stack them in the order they valued them.

Not one of the dozens of participants had Internet-ready features anywhere close to the top of their card stack.

This surprised us. Inside the tech bubble, the Internet was all everyone was talking about. Yet the focus group participants told us the market wasn't yet there. We faced a classic marketing dilemma: market ahead of what customers valued to push the market or market what they told us they prioritized now?

It felt risky at the time, but we decided Internet-ready features should lead our marketing, despite what customers told us. It didn't take long for it to be clear we'd made the right call.

And yet, the exact opposite happened at Loudcloud.

We were the first to offer Internet infrastructure as a service and talk about services in the cloud. We went to analysts and customers and did focus groups to understand sentiment. The closest thing they talked about was a managed service—which we felt wasn't big enough for our vision. We decided we had to try to change minds.

We talked about the cloud being like a power utility that you turned on and used dynamically on-demand. We used the best platform for declaring vision in the world at the time—our co-founder, Marc Andreessen, co-author of the first widely used browser and co-founder of Netscape. We even had a cover story in WIRED magazine.

But it wasn't enough. Back then, not enough people recognized the sea change we saw. There wasn't common language to build on or a groundswell in the industry of others doing similar things. In 2000, the world was struggling to comprehend a cloud service—a ubiquitous concept today. We were too early. It was too big a leap for people's collective mental maps for our messaging to stick. Our attempts to position ourselves in a new category failed.

Whether or not your story is right for right now requires judgment steeped in deep knowledge of customers, the industry, and technology trends. If you're pushing the boundaries of what people believe, you need to show the industry is moving with you. Competition isn't a crowd; it's confirmation. This was what the Internet situation had that the cloud one didn't at that time.

Concrete evidence—industry trends, customer stories, adoption, pundits, data—is at the heart of messaging and positioning feeling "right" at any given time. Let's explore some of the tradeoffs and how to make smart decisions.

Right Category: Create New or Redefine Existing?

Some suggest the only way to play big or to lead a category is to define a new one. I disagree, and Table 24.1 shows why.

The table isn't meant to be comprehensive (and some might disagree with how I've categorized), but you get the gist.

In general, creating a truly new category is *really* hard and demands a lot of persistence, time, and resources. There are far more examples of great success stories from those who redefine

Storyteller

Table 24.1 Category kings exist in both scenarios.

Created new category	Redefined existing category
Amazon (the store)	Google → surpassed Yahoo
Amazon Web Services	Apple iPhone → surpassed Blackberry
Netflix	Facebook → surpassed MySpace, Friendster
RedHat[1]	Salesforce → surpassed Siebel
	Slack[2] → surpassed Hipchat[3]
	Spotify → surpassed Pandora
	SpaceX → now used by NASA
	Microsoft → nearly every business it's in
	Tesla → redefined everything in electric cars

existing categories than of those who defined something truly new. The products created by the giants on the right side of the column were deeply innovative, but the space was understood because of the products that preceded them.

True category creation is not a marketing event. It requires an industry shift and others to follow. To create a new category, you need great product strategy, business strategy, marketing strategy, exceptional leadership, resources, luck, and consistent execution over time. There must be ample evidence you can point to outside of your company—and in your customer's world—that confirms a shift is occurring. There just aren't that many companies who are really good at all of that.

In technology, all categories break into smaller subcategories. The boundaries constantly adjust to new technology, trends, and many new companies jumping in every year.

That's why it's easier to redraw lines around existing categories than it is to create new ones. You have that "You are here" marker from which to anchor conversations.

---

[1] Acquired by IBM.
[2] Acquired by Salesforce.
[3] Acquired by Atlassian.

The job of messaging is to connect with customers and what they feel. Sometimes, a vision of what's possible intrigues them. How your product is a part of bigger sea change can be a way to increase curiosity and excitement.

How much to lean into this "new vision" is unique to your company's stage, market situation, and the resources you can apply to the work. Selling a vision requires telling a story in which your product plays just a part. Stories can't just be puff pieces. They need a lot of other parts to be credible and compelling. It requires creativity but equally important, evidence and restraint.

Whichever path you choose, it must be brought to life with a complete story. It should include industry thought pieces, data, videos, customer stories, and evangelists who your audience finds credible. Creating curiosity about what's possible is a big part of what leaders do to take their place in a category and maintain it.

Leveraging Product Managers

Product managers have the clearest idea of where the product is going and why. They are the holders of the product vision and strategy and so should have a clear sense of how the future intersects with what a product does now.

Product managers can inspire story ideas and also ensure technical accuracy. They know which features meet current customer needs as well as which position the product toward a compelling future.

For example, if artificial intelligence is important to where your category is going, promote how it might be used in the future with your product. But it might not be the lead. Show how AI makes a difference in how specific features are better.

Storyteller

Leveraging Sales

I once worked with a company where the CEO fell in love with a way to talk about the flagship product. The marketing team happened to be in the middle of messaging interviews, which included the head of sales who used his own phrasing of what the company did. Marketing liked it but thought it could use some jazzing up.

Before they decided which was the right direction, they tested the CEO's, marketing's, and the head of sales' messages on the home page—each for a week with a "Learn more" link as the test for engagement.

The head of sales' version outperformed the CEO's by over 1000% and beat marketing's version handily too. Needless to say, the sales leader's was the one they went with. When the marketing team examined why his beat the others, they realized the alternatives relied too much on jargon and felt less authentic.

It's not surprising that sales is a critical ally in developing messaging that works. They have to say it daily in a way that makes customers feel like they're not being sold. Don't hand sales finished messaging; collaborate on it together and validate it with the field. The outcome will always be better.

Leveraging Search Trends and Techniques

In chapter 2, I talked about how search engine optimization (SEO) is its own dedicated specialty. It casts a net that lets people discover your product based on how they search.

It is an important input, but it can't tell you your final messaging. That needs to consider a balance of things, like positioning, the competition, and of-the-moment trends that are better used for campaign messaging.

I do, however, find many techniques using search to be easy, fast gut checks on words, concepts, and categories that should inform messaging.

A few search-based techniques to validate messaging:

- **Search Trends.** It's a great place to see how people across the world use terms associated with your space. Trends are most helpful when comparing terms to one another so you see a term's relative place.
- **Search Journey Testing.** Like any other user-journey test, but this time, you're watching how a user searches for your product and the actions they take based on what they find.
- **Targeted ad buys.** Very important in tests like these is to not just test mild variations of the same theme—you're trying to get bigger directional feedback, which means you need to be testing directions that are clearly different from one another. For example, lead an ad with a description of what you do and compare that to an ad that leads with a problem you solve. Test both against messaging similar to your competitors.
- **Keyword audits.** Usually done by specialists, they examine the universe of terms associated with your web presence and that of your competitors. It guides on key messages that could improve search discoverability.

It's a Balancing Act

What's best to message now and how that shapes your future position depends not just on your product and company stage. It depends on the industry and market dynamics surrounding you. That's why search is so specifically helpful to calibrate your starting point and whether you're moving toward your positioning over time.

Storyteller

The earlier in a product's life you are, the more you must tell a story using the pain and problems that make you necessary. Create mental connectors to help people understand what you do. It's also the time to define or redefine how the world sees your category.

The more mature you are, the more understood your product and category is. Product messaging can be more aspirational in mature categories. It's equally important to focus on building toward a longer-term position that redraws the goalposts of your category and leans them toward your company's vision.

When it comes to messaging and positioning, creating a strong story and messaging is just the start. Next comes the challenge of getting everyone on the same page about what to say. That's where the one-sheet messaging canvas comes in.

Storyteller

Chapter 25

The One-Sheet Messaging Canvas

It was a chilly fall day when the CEO emailed leaders from marketing, sales, and product, urging them to come to an important meeting. They were getting beaten in head-to-head deals by a fierce competitor that told a much better story. Their company needed a better pitch *fast*.

At the meeting, the head of marketing chimed in with messaging that was working well in email campaigns. The head of product suggested a catchy tagline she felt captured the promise of the product. The sales managers didn't love any of the new ideas but didn't have any better suggestions. They reminded the group what was working well in the current pitch.

The product marketer was an individual contributor in a process filled with more senior leaders. Because all the input came from leaders above his level, he felt their ideas had tacit approval. He did his best to smooth out transitions between ideas, but one by one, he integrated them into the beginning, middle, and end of the existing pitch. In an effort to rush to get the new story out, the presentation wasn't tested in front of any customers or used in an actual pitch.

The end result was tolerable but not good.

This collaboration-to-mediocrity is all too common. Coming up with great messaging and the story to embed it in is a team sport, but it is product marketing's job to collect the inputs and craft a strong output. They are product managers to the product's story and how it gets messaged.

As I've talked about throughout part 4, strong messaging comes from a strong process. This chapter's One-Sheet Messaging Canvas was designed as the collection point for that process and becomes the artifact for final messaging when done.

When completed, the canvas gives anyone talking about the product an easy toolkit from which to build marketing messages that position and reinforce a product's broader narrative.

How It Works

The One-Sheet Messaging Canvas separates messaging elements into individual building blocks. It makes each part—positioning, customer benefits, how a product does what it claims, and proof those claims are true—individual tools that can be used for any go-to-market purpose. You don't use all of them all the time, just what you need at the right time.

When the canvas is finalized, everything in it builds the case for why people should believe a product is a good one. The product's market position should also be clear.

Don't fill it out like a form (Figure 25.1). At the start, document ideas worth testing. Then test, probe, and let it be a hot mess while ideas are refined with teams. Put them in front of customers in the mediums in which they'll see them. The process is iterative, and like any good discovery process, be prepared to throw away as many ideas as you keep.

The completed canvas becomes the primary written artifact for all teams talking, writing, or creating something about the product. By having everyone use the same toolkit, the consistent messaging over time reinforces market position.

Here's how you use it.

> **Setup: Messaging discovery. Open-ended questions to find customer language and context.** These are questions that can be asked as part of market fit discovery

One-Sheet Messaging Canvas

Positioning Statement Clear, Authentic, Simple, Tested	What Product X does		
Support / Key Benefits	*Defensible customer benefit...*	*Or key area of strategic differentiation...*	*Or qualities important to customer*
Customer Segment Key ● *Primary Decision-Maker* ▲ *Secondary Decision-Maker* ◆ *Technical Influencer*	*Why something is a benefit to customer...* ◆ Benefit 1 ● Benefit 2	*Or why something is of value to customer...* ▲ Benefit 3 ●▲◆ Benefit 4	*Or why something is important to customer.* ●▲◆ Benefit 5 ▲◆ Benefit 6
Proof of Value: Evidence to support above claims			
Business Conditions *What to look for to know customer is good match for product*	• Proof Point 1 • Proof Point 2	• Proof Point 3 • Proof Point 4	• Proof Point 5 • Proof Point 6

Figure 25.1 The blank One-Sheet Messaging Canvas that starts the process.

Storyteller

work: How might they describe this to a friend? What problems do they experience that make them look for a better way of doing things?

Listen to the words used and how they talk about their problems. I'm also a fan of micro-surveys, short, to the point, one or two questions, always open-ended. They capture what people think without any prompting. Questions like "Why didn't you try our service?" or "What can we do better to meet your needs?" reveal what messaging must clarify.

Step 1: Decide on your most important Customer Segments and shape messaging around them. Your messaging is not for all your audiences, just the most important ones. Make it clear for them. There are plenty of other tools to reach additional audiences. This is how you avoid trying to say all things to all people and mean nothing.

Step 2: Come up with "starter" messaging and key support messages. Think about what customers want to hear. What makes you credible and intriguing? Maybe it's a broad product benefit, a novel way your product does something, an ecosystem you're associated with, or technologies you unlock. Look at all prior iterations for inspiration, and don't fall into the trap of just saying what you want to say. These pillars create the frame into which everything else falls.

Step 3: List areas of value in customer friendly language under each pillar. Don't just list features and what they do. Tie what a feature does to its benefits or use cases. For example, you may explain a benefit is possible because of a unique way your product operates. What winds up here is very specific to audiences.

Step 4: Designate messages as appropriate for specific audiences. Not every message is appropriate for all customer segments. Identify a unique symbol for each segment the canvas is for. Put that symbol in front of the messages that are appropriate for that audience. For example, if you're targeting a developer audience and designated their symbol a square, you might have a square symbol next to APIs that support data integrations. Whereas an increase in developer productivity has both a square and a circle, the latter being equally meaningful to engineering leaders. The same messages can be used for multiple segments.

Step 5: Provide evidence. Messaging is just as much about what others say about your product as what you say about it. Think about what enables evangelism. Customer stories are easier for people to remember and share. Evidence can be uses cases, data about the product being better, research or analyst quotes or proven ROI. What matters is that this section feels full of facts and makes anyone using the evidence feel that it's credible.

Step 6: Test it with customers in the intended mediums (email, website, presentation). If you have a sales force, start there. They can immediately use it and gauge customer reactions. Listen to responses. You're not looking for customers saying "This is great!" Rather you're looking for anything that drives curiosity and deepens conversation. If they don't say anything indicating interest, keep working the messages. You want to strike an emotional chord, discover third rails to stay away from, or find yellow-brick roads to follow.

Step 7: Refine. Remember Netflix used "cancel anytime"? What works might not be related to your product. Take all you learn from step 6 and refine what's in the canvas. By this step, you may have landed on a tagline that's working. Only include it in the canvas if you want others using it and you're trying to drive consistency.

One of the best ways to refine the canvas is to practice what's on it out loud. Try to listen as someone new to your product would hear it. You're listening for overly jargonistic words and when things feel too forced. Do a final check with the CAST guidelines to make sure the messaging does what it needs:

1. **C**lear. Is what you do easy to understand? Does it drive curiosity? Is it clear versus comprehensive?
2. **A**uthentic. Does this sound like it's from the customers' point of view? Will they feel known? Is it authentic versus authoritative?
3. **S**imple. Is it easy to understand what's different and compelling?
4. **T**ested. Was it tested in the context customers experience it?

The Messaging Canvas in Action

Figure 25.2 is an early draft from a startup called IndexTank that was eventually acquired by LinkedIn. They offered search as a service, and their targets were developers and operations leads at large websites using search. In their case, each pillar supported a different key audience.

During their messaging discovery work, they found a surprising third segment: the developer evangelist. This audience didn't plan to use the service themselves, but they liked testing new tech and were highly engaged in developer forums. Because many developers prefer learning via word-of-mouth, they realized it was important for their messaging to make general app developers—not just developers working on search—curious too.

But after testing on the website, they shifted focus again. The pillars remained the same, but the short version of the positioning statement became "Easily add powerful search to your site. Good search is good for business." This was because those curious app developers weren't paying customers. Better search enabled more revenue for larger websites, and these made them the better customers for IndexTank's business.

Positioning Statement
Clear, Authentic, Simple, Tested

IndexTank is a powerful search API that lets you easily add real-time, customizable search to your apps.
Short: Easily add search to your apps.

Support / Key Benefits	*Powerful API for today's apps.*	*Custom search that you control.*	*Easy, fast, and hosted.*
	Value to Customer		
Customer Segment Key	● **Real-time:** true real-time indexing means updated results in real time	●▲ **Fast:** API runs in memory, not storage disks, so gets results in fractions of seconds	● **Scalable:** supports 45MM searches and counting
● Ops/IT person who did search	● **Geo-aware:** latitude and longitude of person submitting search can be used to determine search results	◆ **Stop hiding search:** let users dive into your content because you're confident it meets their needs	●▲ **Free** 1st 100K documents so no risk to start
▲ Freelance developer b/w jobs or brushing up skills	● **Social:** Uses votes, ratings comments, likes, or page views for results	●▲ **Fuzzy:** Show people what they mean, not what they type. Supports fuzzy syntax or partial inputs.	●▲ **Easier and faster** to add key search features than SOLR, Sphinx, and Lucene
◆ Product person who wants better search for users	●▲ **Mobile-ready:** call API directly using our Java client	●▲ **Autocomplete:** Show results faster with less work for users	▲◆ Ruby, Rails, Python, Java and PHP all supported
		●▲ **Snippets:** Let users preview search results without having to click	
		● **Faceted:** Let users dynamically control results displayed	
Business Conditions	**Proof of Value**		
· Search important or a big problem for their business	· **Reddit**—became #1 social news site within a month of improving their search of 10MM docs daily	· **Blip.tv**—weighted results important for business model	· Sample data set can be "live" with queries in 2 minutes
· Need more real-time or social filters to make results relevant to today's customer	· **Twitvid**—search is how to find content so recency, social ratings, and real-time are essential	· **TaskRabbit**—smart relevance and fuzzy logic critical for finding related tasks	· Other platforms require much more time, work, and customization to equal built-in features
· Removes search as limitation for UX or customer success		· **Gazaro**—faceting allows users control they expect and want	

Figure 25.2 Early draft One-Sheet Messaging Canvas from IndexTank.

The positioning statement shifted to what their most important customers needed to hear.

The process isn't hard but requires some discipline for product marketing to not "close" on messaging before it's been tested. Just as important are the deliberate decisions on how to apply learnings to final messaging. It might not be the best performing message. It might be what's better for the business and positioning.

Answers to Most Frequently Asked Questions

How long does this whole process take?

For startups, messaging changes a lot. The first pass that gets iterated on should come together very quickly (less than a week), but the test and refinement process can last a month. Updates to messaging might happen frequently.

For more mature companies, the test and refinement period might take longer—it could take months depending on how deeply they explore and test—but once set, it tends to stay set for the life cycle of a major release.

How much should customer input drive messaging?

A lot, but messaging is customer informed, not customer driven. Customer insights are inputs. Only you have all the pieces around them + market + technology + business needs.

How many should I do if we're a multiproduct company?

One for each product or product family.

What if it goes beyond one sheet?

Part of a single-page canvas is to force constraint and prioritization. It pushes you to trade off what's most important versus the tendency to be comprehensive. There are other tools in the toolbox (the website, white papers, feature comparison charts) that can cover deeper information needs. The purpose of the messaging one-sheet is so the most important messages get reinforced consistently by everyone that uses product messaging to do their jobs. Concision matters.

Part Five

Advanced Product Marketing and Leadership: How to Do and Lead It Better at Any Stage Company

Chapter 26

Leading and Transforming Product Marketing

When Mala Sharma arrived in Silicon Valley in 1996 to start her career in high tech product marketing, she was shocked. No one knew daily or weekly sales figures. There was no accountability between a market action and its business impact. The metric for success for product marketing (or marketing for that matter) seemed to be: if product collateral was getting done, everything was good enough.

Earlier in her career, Mala worked as a brand manager for consumer-packaged goods with Unilever in India, where she was trained to act as a virtual general manager. She knew her product's weekly sales figures, the costs for everything going into them, and the key drivers that made them work as a business.

In Silicon Valley, however, she realized no one around her had any idea of the strategic opportunities product marketing could unlock simply by better understanding customer sentiment, competitors, and markets. In her first product marketing role, she immediately kicked off both quantitative and qualitative customer research and found a gamer niche that changed the packaging, pricing, and go-to-market strategy. The product's success changed how her company thought about the role.

It strengthened her conviction to shift the perception of the job wherever she worked. When she arrived at Adobe as the Director of Product Marketing for Photoshop, the product managers figured out the features, then handed those plans over to product marketing to work on messaging and enabling sales. So Mala started showing up at every meeting she could to make sure she understood the market opportunity better than anyone else. How do we grow this business? Is pricing an issue? What's the product's go-to-market? She modeled what strategic product marketing looks like in every meeting and started shifting perceptions.

Her responsibilities soon grew to include all of Adobe's Creative Solutions, and she continued to transform the function by hiring leaders from the outside. Unlike many of her peers in Silicon Valley, she didn't have a bias toward technical backgrounds. Instead, she looked for a combination of consumer packaged goods, technology, and/or management consulting experience. She held the bar high, realizing that while it was hard work to find them, the combination wasn't as unicorn as most people thought. The blend provided a deeper pool of people with proven strategic acumen and success as business-minded operators.

Adobe's annual revenue exceeds $11 billion, which means the size and complexity of the organization is high. Mala had to make sure the role of product marketing was clearly defined relative to product management, campaign marketing, and the go-to-market teams that work more closely with sales. It's why she enlisted the help of the software life-cycle team—a joint software program management office—to agree on the high impact points between product and marketing. They created written artifacts to ensure the right work got done and could be communicated to all stakeholders. Equally important, she invested in training, so as new product marketers were brought on board, they immediately knew the model for product marketing and what was expected of them.

Her most important insight in all these transformations? Product marketers succeed when they are willing to challenge a conversation based on customer insights. They, of course, need the data, market and customer insights to do that. They must be unafraid to put difficult choices on the table for teams with whom they collaborate. This, in turn, constantly improves product go-to-market thinking and empowers product marketing to find new ways to unlock business growth.

Mala had a clear vision of what product marketing could do because she'd experienced its power in prior roles. It made it easier for her to envision how it should operate and model the behavior for others to see.

Most aren't so lucky. That's why it's important for leaders to clearly define how they want the role to operate and communicate this across the entire organization. The four fundamentals—ambassador, strategist, storyteller, and evangelist—anchor the role, but how they work in a company's operating environment is scoped by the managers leading them.

Starting with how product marketing is organized.

Where Should Product Marketing Report?

As it relates to product marketing being effective, how it is allocated to cross-functional product teams is more important than where it reports. I talk in depth about specific recommendations in Chapter 8, so I won't belabor it here.

According to Forrester,[1] there is no typical ratio. But based on their data, the average ratio is 2.6 product managers to one product marketer and goes up to 5 product managers for every product marketer.

PM to PMM ratios aside, whether or not product marketing should report to marketing or product is one of the most common questions leaders grapple with. What's best depends on two major factors:

[1] https://www.forrester.com/blogs/whats-the-right-ratio-for-product-or-solution-success/.

Advanced PMM

Factor 1: What Business Problems Are You Trying to Solve?

Organize under marketing. When a product is well established in its market, growth requires a lot of market focus. Customer segmentation gets more refined (ex: micro-segmentation) as more obvious segments have already adopted. Marketing strategies—like partnerships, product suites—and much more GTM coordination means product marketing does well as part of the marketing organization.

It lets product marketing act as an overlay *across* products. Markets and customers drive go-to-market thinking, not individual products. Products are a portfolio from which a company can grow revenue. Product marketing organizes by how customers experience that value. For example, is it experienced differently by a large enterprise versus an individual buyer?

Organize under product. In a world of continuous development, for highly technical products, or if the company struggles to communicate well about its product, product marketing often closes this gap better when reporting into product. This reduces the friction for product marketers to get better product information to go-to-market teams.

It also makes it easier for product marketers to be the voice of the market to product teams, as they're more integrated into the product organization. This reporting structure has the added benefit of helping increase product teams' market acumen.

Factor 2: Which Leader Has the Capacity to Help the Function Reach Its Potential?

If you're lucky enough to have an extraordinary leader with a unique combination of superpowers—either their experience or leadership capabilities span multiple areas—take advantage.

For example, you might have an extremely market-savvy chief product officer (CPO) who sees things more strategically than her chief marketing officer (CMO) counterpart. Put product marketing underneath that CPO, even if your markets are mature. The credibility lent to the function by reporting to an extraordinary leader is part of what helps the function succeed.

Conversely, if you have a CMO with deep credibility with product leadership, this is a way to have leadership model a healthy collaboration between go-to-market and product functions. Have product marketing report to that CMO.

Defining the Scope of the Role

Product marketing is full of versatile generalists. This can make defining the boundaries of the role tricky.

When a product is early in its life cycle, the generalist is very valuable. A lot of interpretation and adaptation is needed to discover a product's go-to-market. A generalist can broadly apply skills in discovery or hands-on marketing work.

This shifts, however, as companies mature and a product's repeatable go-to-market is more clear. That's when product marketers tend to specialize around vertical markets, marketing or distribution channels, or customer segments.

The size of the product marketing team likewise tends to grow over time as products become more successful and their markets more complex. You might have a product marketing team for a single product that has a group of specialists driving each aspect of core go-to-market, plus another focused on a major vertical market (e.g. accoutants) and others focused on each major customer segment (e.g. enterprise).

Define the role's scope by what is most important for where the business is going and how the team is organized. Then communicate it clearly to teams outside of product marketing.

If you want to bite off bigger product marketing organizational change, don't start with a full-scale reorganization. Pair your very best product leader with your very best product marketing leader in the most important market segment(s) for your company. Have the teams try out the model and tools. Adjust for your company's unique situation, and then roll out a version that works within your organization's reality. Doing it this way gives a working reference point for everyone to understand how product marketing is intended to function.

In the case of a product marketing reboot—redefining the charter of an existing product marketing group—take the time to examine the gap in expectations and performance from its partner functions. Then create a new structure and process that close those gaps. Put in place clear metrics that help everyone see if the old problems are getting solved with the new organizational structure and charter.

Let's explore some of the organizational dynamics and fixes to consider when revisiting a product marketing team's charter.

Product Marketing and Product Management

Trust in product marketing's knowledge of the customer and market is foundational for this relationship to work well. It's often little things—a lack of curiosity about the product, data not used in discussions, or marketing needs inserted into product discussions seeming random—that undermine trust.

Finding these issues may require observing interactions first-hand—such as attending team meetings where the product marketer works with their product manager counterpart to observe how they interact. I also recommend having direct conversations with product counterparts to diagnose issues.

Conversely, a lot of product managers don't understand just how hard it is to create a product's story, its messaging, or go-to-market. For them, the difference between one choice of words or a particular activity versus another can feel arbitrary.

When defining roles to fix these dynamics, focus as much on shifts in processes as well as clearer expectations of what "good" looks like for both sides.

Product Marketing and Go-to-Market Teams

No go-to-market plan survives contact with its customers. You can't predict the actions of others, and product marketing must direct its response.

Frustrations arise when product marketers focus on the execution of a plan more than reacting to what's actually happening in the market. If either sales or marketing feel what they've been given by product marketing isn't working well, they need clear ways to discuss improvements.

When revisiting roles or responsibilities, make sure standard processes get defined for how priorities get decided between teams. For example, weekly meetings where sales can impact how quickly they get a new sales tool or shift the plan for upcoming events.

Product Marketing and Executive Leadership

Most executives don't have a clear idea of what product marketing does or should do. Product marketing leaders need to define what their team does relative to company priorities and connect the go-to-market actions driven by their teams.

This starts with a clear product go-to-market plan and then more detail on how specific activities and plans add up and support business goals. Likewise, they should say how the team's activities should be measured both in the shorter (quarterly) and longer (annual) terms.

The Importance of Inclusive Team Norms

One of the most important roles of a team leader is making sure the team's composition sets it up for success. This is not just about hiring people of different genders or ethnic backgrounds. It's the ability of a team to see things they might not otherwise see but deeply matter.

Advanced PMM

If a product is available on the Internet, it's instantly global even if priced for US markets. If a product is more popular with men than women, the answer to broaden appeal isn't "let's do a pink breast cancer version." If a product asks for personal information, some customers might object.

These were all real situations faced by product teams where the dominant opinion of the team was swayed by the presence of someone with a different perspective. In most cases, that person had a different ethnic or gender background than the rest of the team. In each case, the teams made better decisions than had that person not voiced their point of view.

The benefits of more diverse teams are widely known; they outperform their less-diverse counterparts. But many people don't realize it goes beyond making diverse hires.

In 2012, Google embarked on a study, Project Aristotle, that examined hundreds of their teams over a year to understand why some struggled while others soared.

They found performance was less about who is on the team and more about how the team worked together. The best and most effective teams enabled a foundation of trust so there was no fear of conflict and unfiltered, passionate debate of ideas flowed freely.

Here were the most important factors that had a significant impact on outcomes and correlated to consistently better performance:

- *Psychological safety.* This is having the confidence that it is safe to take risks with asking questions, offering new ideas, disagreeing, or sharing something personal.
- *Dependability.* Team members all reliably complete quality work on time and don't shy away from being held accountable.
- *Structure and clarity.* This is simply a clear understanding of job expectations and what it takes to fulfill them.

- *Meaning.* Having purpose in one's work. Meaning is also personal.
- *Impact.* People want to feel like they're making a difference and that their contribution matters.

Team dysfunction stems from group norms born out of habits in team dynamics. Sometimes, we don't even recognize a team isn't operating optimally.

This matters a lot for product marketers because they operate so cross-functionally. They must feel confident to influence outcomes with teams where they don't share a reporting structure. They must feel safe disagreeing.

It's the leader's role to make sure this type of working environment is in place.

The most effective leaders model the desired behaviors, like making quiet people feel heard, speaking last instead of first to ensure others share their opinions, embracing different points of view even if you don't agree, and not defending against opinions you disagree with. Show candor, respect, clarity, and a willingness to have hard conversations whenever necessary.

When leading product marketing, examine not just the business goals and functions of the role. Look at who sits around a team's table and how the team functions to enable great performance.

Chapter 27

How to Hire Strong Product Marketing Talent

I was recently asked to sit in on final interviews of three product marketing candidates. All of them said a lot of the right things—focusing on strategy, messaging, knowing the customer—but I only recommended one as a hire.

She was not the one with the most polished presentation. Her strengths lay in skills that are harder to teach—the humility to admit what she didn't know, the ability to adapt her plans in real time, and a genuine curiosity about the product. The other two said they had those skills; only the third showed me she had them.

This is not to say the other candidates didn't have impressive strengths. One was exceptional at verbal communication and had a strong understanding of the role. The other was great at navigating organizational hurdles and had in-depth management frameworks. But the company doing the hiring was an early-stage startup. Adaptability and curiosity were the more important skills for their hire.

This chapter is for leaders who want to learn how to hire great product marketing talent.

Assessing the Skill Set

Everyone wants to hire unicorns—mythical product marketers who have it all and can bend spoons with their minds. They exist, but they are rare.

More common is to optimize a hire around what's most important for your company's stage and how much you can grow or develop skills. This is not an issue of company size. A great manager/mentor and teammates who use strong frameworks are possible at any stage.

Inevitably, you're making some tradeoffs. I'll use the three product marketing candidates (the curious, the communicator, the collaborator) to reveal what these skills might look like in real life.

I'm using the chapter 7 skills to frame what to look for and then I'll offer some questions that assess those skills.

- **Deep customer curiosity, strong active listening.** Being great at product marketing doesn't mean you have all the right answers. It means you're great at listening for market signal from anywhere: customers, product teams, sales teams, the news. Candidates need to be good at making connections with vastly different information and sources. They then must have the ability to translate them into thoughtful actions and messaging.

   Evidence: Ability to adapt point-of-view based on changing situations or information. For customer-centric messaging, review samples of their work.

   Interview question: Can you talk through a time when you had a plan but changed it based on something new you learned? Push back on any assumption the candidate has made about something they've told you—do they reconsider their point-of-view?

   This is where the curious candidate beat out the other two. Whenever pressed on an idea or plan, she didn't hesitate to adapt it or say, "I'd lean a lot on other experts in the company for help." This shows a learning instinct for problem solving. The communicator and collaborator didn't exhibit any tendency to diverge from their prepared plans.

- **Product curiosity and technical competence.** Products change a lot. Good product marketers show evidence of quick adaptation and a growth mindset. They constantly develop their skills and genuinely enjoy products. Beware of being seduced by industry expertise. It can be a double-edged sword—great knowledge of a space can mean a fixed mindset about what should be done.

  Evidence: Do they talk about products, unprompted, in the interview? Did they choose different jobs and industries over the course of their career? Do they exhibit sincere interest in the category or the product?

  Interview question: What were you looking for when you jumped from [company X] to [company Y]? What lesson(s) did you apply from [insert job] into how you did your next job? What product do you love right now?

  The curious candidate showed how she looked at the space—including how she might examine the competition—even though she was not an expert. The other two had product knowledge as a checkbox in their initial plans for how to get up to speed.

- **Strong on strategy *and* execution *and* business savvy.** A lot is revealed in what someone chooses to discuss as career highlights, such as business results tied to their accomplishments. You're looking for how they define excellence and a thoughtful approach that connects to impactful outcomes.

  Evidence: Great business results in a reasonable timeframe. A collaborative approach that leveraged many on a team. When discussing product launches, do they map it back to what it did for the business?

Interview question: What do you consider your biggest success so far? How did you know it was a success?

All three candidates displayed evidence of thoughtful plans and how they would manage them through organizations.

- **Collaborative.** Look for systems, not just relationships, that drive results. If they worked often with sales, did they collaboratively develop sales tools? Ask directly if they've developed messaging together with product and what they thought of that process. There isn't a single right way—you're listening for whether or not they let inspiration come from collaboration.

  Evidence: Discussions of teams, processes that elevate or accelerate success.

  Interview question: What is the best product/sales/product marketing relationship you've seen in action? What made it good?

  The collaborator was the strongest candidate here, showing evidence of many product launches where she had to manage tricky situations across many functions. This meant she was well suited for a more mature organization where so much of succeeding in the role is corralling the necessary go-to-market functions and getting them to act as one.

- **Strong written and verbal communication** comes through how they communicate during interviews, email, or even in their resume or LinkedIn. Strong verbal skills don't always translate to strong written skills. Someone can command a room but have flabby writing or vice versa.

  Evidence: Outside of interviews, use exercises in the interview process to probe these skills. Consider having candidates do a presentation. The very best presenters will not just be engaging; they will make the presentation interactive. It's always good to look at examples of the candidate's work.

Advanced PMM

The communicator was very crisp in her talk track and stopped to check-in with her audience throughout. While smooth, she was not as strong if we asked questions that moved her off her talk track. It became clear she was really good at what she had prepared, not as strong reacting on the fly. This super communicator skill set is great at the growth stage where clear, consistent messages helps the company scale.

How to Assess Raw Ability versus Experience

In my nearly 30 years of doing interviews, there is one question I've asked every product marketing candidate—from individual contributors all the way to executives. It continues to be an effective way to separate dynamic thinkers with a more intrinsic talent for marketing versus people who are simply good at executing a more formulaic approach.

It starts with this: *"Tell me about a product or company you think is doing really great marketing. It can be anything. And tell me why you think it's great."*

The question is intended to create a roughly 10-minute conversation that reveals how broadly a candidate defines marketing success and the diversity of their tool set. This approach requires you to work just as hard as the person you're interviewing because there isn't a right answer; it's truly a discussion. The setup starts by focusing on what one company is doing particularly well.

But the real test is the corollary of your opening question: *"Now pretend you are now a marketing leader at a competitor to the 'great marketing' company. What two or three things would you consider most important to do in order to compete with [insert their first named company] well?"*

The reason to discuss companies other than your own is because neither of you has an information advantage. It forces you to pay attention to their thinking at least as much as their answers. It also doesn't create bias in what you're listening for because you can't have a set expectation around their answer.

Here's what you're looking for in a great candidate response.

- **Breadth of marketing tool set.** It is crucial to push the candidate to talk about *why* anything is good. This reveals how broadly they define marketing and their *understanding of marketing levers*—do they bring in brand, customers, product, clever campaigns, and articulate why anything is effective?

  If a candidate doesn't talk much beyond an advertising campaign or can't get specific about why they think a product is better, they've failed. If the candidate only goes through the four P's (product, price, promotion, place) without any original thinking, I consider this a fail. It suggests a person relies on a formula instead of really tuning into what makes any marketing idea connect with customers or a market.

  I always give them multiple chances using a technique called breadcrumbing. You suggest some examples or introduce new data to lead them a bit and see what they do with it.

- **How do they deal with rapid change?** Technology trends change quickly. When you introduce marketing for a competitor, this really separates candidates. The ability to think on your feet and come up with new ideas on the fly is essential for great product marketing because every category in tech is dynamic and rapidly shifting. A great product marketer rolls with the changing landscape.

- **Are they open to new information?** You give the candidate every opportunity to succeed, which includes introducing more information to help keep the thinking going. For example, shift the target customer or any other market factor. Does that change anything they do?

Advanced PMM

Listen for assumptions they've made and see how the candidate handles unexpected shifts.

- **How do they manage constraints?** I sometimes add a budget constraint into the discussion. "If you only had a $250,000 budget, would it change what you do and how you prioritize?"

Let Every Candidate Shine

You might wonder what's wrong with the typical "What are three things you think we could do better in our marketing?" interview question. You know a lot more about your company than the interviewee and probably already have some ideas you think are pretty good. It is hard not to measure the candidates' answers against your own. If marketing isn't your expertise, it's especially hard to recognize if an innovative idea is superior to what you already believe.

Discussing companies where neither of you knows the answers means you'll be less apt to rush to judgment when ideas are different from your own.

For calibration: at the entry-level stage (straight out of Master's or MBA program or an exceptional undergrad) about one or two in ten are dynamic marketing thinkers. The ratio improves to more like one in five, sometimes one in three when you're working with a really good recruiter for director level or above product marketers.

Hiring well is the most impactful thing leaders can do for product marketing. It's worth being rigorous in how you interview and assess. But hiring is just the start. You'll want to develop talent to get the most out of anyone you hire.

Chapter 28

How to Guide a Product Marketing Career

A career path in product marketing can go in almost any direction—leading growth, product, marketing, or a business unit. There is no set track. Strong leaders help product marketers become versatile and highly skilled so any career path is possible.

This chapter provides a roadmap for how to guide product marketing skills over time. The year ranges should only be used as guides.

Early Career: One to Five-ish Years

Broad, rapid learning is the name of the early career game. Hone their ability to interpret market signals in addition to learning how to do specific functional tasks well.

Don't shy away from detailed and concrete feedback. Commenting on an email's content is not micromanaging if it makes all subsequent emails more effective. Help them learn what excellence looks like.

It's a great time to let them lead initiatives, such as a website redesign or crafting strategy at a campaign level. Make sure they learn how to measure impact and success.

Functional Skills
- Ability to interpret customer and market research
- Able to do insightful market tests and customer interviews
- Competitive analysis
- Product demonstrations
- Sales tools
- Website content
- Thought leadership content

Foundational Skills
- Writing—strive for more concision and story telling
- Oral—learning to read the audience and adjust accordingly and on the fly if necessary
- Holding productive discussions with customers, sales, and product

It's equally important in these early years to give product marketers frameworks and tools that let them be more systematic in how they approach the job. It gets them more efficient instead of recreating the wheel. Whenever possible, provide examples they can model their work after.

Early career product marketers are ready for the next level when they are successful at managing someone (summer interns count!) or managing complex, high-impact projects that very visibly move the business forward. That last nuance is important; doing something great just within the marketing team is not the same.

Mid-Level: Five to Twelve-ish Years

This is where functional skills expand as well as the expectations to guide and lead across functions grows. Expect to see some expansion in how people think about their job (should we partner with a company?) and the ability to create with much less guidance.

Director-level leaders are more visible as leaders across the company. They lead major launches. They may be put in charge of multiple product lines or lead the transition from product to solution marketing, crossing multiple products and market segments. The level of complexity for success increases.

Additional Functional Skills
- Deepening knowledge of marketing specialties, such as brand, communications, digital, demand generation.
- How to enable partners
- How to market to verticals

Additional Foundational Skills
- Able to lead campaigns across multiple functions
- Excellent at communicating well with and commanding respect of product and sales peers
- Able to hire well
- Management—knows how to lead a team and scale through others

The best leaders at this phase of a career can teach key marketing skills to others and make team members feel productive. Results eventually come from leading others and less from the individual herself. Although at the Director level, the person is still a highly effective doer too.

You know a person is at this stage when they get requested—by name—to be a part of complex or important cross-functional projects. Don't hesitate encouraging movement to lead other functions at this stage—like product management or a function in marketing. It will only make them better at whatever job they do next.

Senior: 10+ years

Great leaders are evident through more than just having done their jobs for many years. They show success and ongoing

Advanced PMM

evidence of learning in a broad range of scenarios. They are also versatile, not just good at applying a familiar playbook.

Senior people should have seen many product launches and maintenance phases of products. They should be good at setting up processes and systems that let their teams perform better, especially with other functions.

They should be able to act as swiftly as markets demand.

They should also have experienced failure and be able to talk about what they learned and do differently as a result. If they can't discuss failure openly, it means they haven't stretched or lack personal insight. This last point often separates someone who is still mid-career from someone who is truly senior and ready to lead a function.

Additional Functional Skills
- Company spokesperson
- Great at negotiating challenges or conflict across cross-functional teams

Additional Foundational Skills
- Leads together with other functions—believes their success is tied equally to others
- Is seen as a leader by other functions

The hardest jump in this stage is from director to VP, largely because the skills aren't as much functional as they are leadership soft skills. Years ago, I published a blog on this that got widely read even in non-tech companies. I'm including the full text here.

I'm a Great Director of Marketing. Why Am I Not a VP?

Years ago, I watched a talented director of marketing take a high-profile startup from stealth to launch to millions in revenue.

She built out product marketing, corporate communications, PR, brand, and partner marketing teams. In all, her marketing organization grew to nearly 20 people in less than a year, including three experienced directors who all really liked working with her. Looking around, she saw others with less experience in their functions with VP titles and she wondered, *Why am I not one*?

She went to her CEO and asked why, despite a long list of marketing accomplishments, she wasn't a vice president? His answer was, "I'm sorry, but you're just not quite ready." Not only was she hurt, she couldn't understand why, and he wasn't able to clearly explain his reasoning. It was frustrating and demoralizing. And she left that exchange not knowing what she needed to do to get there.

That director was me. And that CEO was Ben Horowitz.

What I couldn't see then, I see very clearly now because I've worked with or been on the interview path for hundreds of marketing leaders and recognize the symptoms I exhibited myself.

I'm sharing what I wish Ben had told me then.

These lessons don't just apply to directors wanting to level up. They apply to any marketing leader who is functionally excellent but feels like their abilities aren't quite reflected in their title.

Strap in. The work might be different than you think.

Stop focusing on marketing excellence. This is counterintuitive but is actually the biggest tell between someone who is functionally excellent and someone who is truly ready to assume a meaningful leadership position. Marketing leadership is not about individual marketing abilities—it is about your team's ability to perform and your ability to enable that performance. A VP must create connective tissue between people and functions, providing an environment where people are inspired and can grow. If your outlook on marketing excellence is still about what you've directed and not what you've created structurally for your *team* to be excellent, you've got work to do.

The company > team. While related to the above, it's not the same. Often I see this in the form of a marketing leader saying his team did a great job on x or y. It comes across as either promoting or protecting the team, both of which are important, but at higher levels of leadership, the "we" needs to be cross-functional. Are you defining excellence within the company's broader goals and multiple functions' abilities to succeed? Are you attuned to leadership dynamics among your peers, and how do you navigate them? The latter is some of the most important work for senior marketing leaders, so make sure

(*continued*)

Advanced PMM

to forge effective partnerships with sales and product and define your success together.

Redirect to "why" versus "what." This can feel tough because everyone has an opinion about marketing, but few actually understand how it works. Marketing teams tend to focus on showing everything the team is doing and sharing campaign results or MQL metrics. "See what we're doing? It works!" These metrics mean something to marketing people but not much to the rest of the organization. Great marketing leaders take a step back to not just plan and report. They spend at least as much time helping the organization understand the why. This work is hard because there is no reward for it, nor is anyone asking you to do it. But it is what ultimately makes the rest of the company feel like the function is being led versus just being done.

Don't be an expert, be open. Let's be honest. There is a lot of talk about embracing mistakes, failing fast, and showing vulnerability. But expertise is both rewarded and embraced, and often essential to being viewed as credible. How do you balance this? The marketing leader finds ways to communicate expertise while still showing openness toward others, inviting participation.

This balance is serious Jedi-level $h#t (I'm still working on it). You're never really a master. But the difference between a "functional expert" (director) and "leader" (VP or CMO) lies in tone, tenor, and self-awareness. You don't have to be perfect, but you do have to develop tools to navigate challenges with grace. Experts feel closed. Leaders feel open.

The hardest part about all this? Getting quality feedback so you know what to work on. So much of this is subjective and not about marketing skills. If you find yourself butting up against a title ceiling, ask for brutally honest feedback from people who want to see you succeed. Then forge a plan with a peer, coach, mentor, or manager. And if no one inside your company can do it, that's why career coaches exist.

Better yet, don't wait for your next annual review or lack of a promotion to work on some of this. These are essential leadership skills at every level, and learning how to practice them will benefit you at any stage of your career.

Product marketing is truly a great on-ramp for any career path in tech. It can lead anywhere, and if you're lucky enough to lead people doing it, help grow them into tomorrow's company leaders.

Chapter 29

Product Marketing by Stage

Early, Growth, Mature

Los Angeles is car country, which is why Michelle Denogean, a seasoned car industry Chief Marketing Officer, calls it home. But she was willing to make the long commute to San Francisco because Roadster—a startup that helped car dealers sell online—landed the country's largest dealerships as customers and was in hyper-growth mode.

The product team was filled with veteran engineers. They'd "been there, done that" at multiple companies together and were highly productive. But it meant the go-to-market teams felt like outsiders, disconnected from what was being built.

Michelle wanted to close this gap with product marketers. She made sure the product team interviewed all the candidates since they'd be working so closely together.

By Michelle's standards, the quality of candidates was high. All of them had years of deep car industry and product marketing experience.

But the product teams kept passing on candidates, citing "not deep enough on product."

By the third rejection, she pushed the team harder on why they kept saying no. That's when she learned the product team was looking for something totally different: a person to help Roadster's customers use the product more in their day-to-day work.

231

It was a complete surprise to Michelle, who felt such work was more suited to either a product manager focused on product engagement or customer success manager, not a product marketer. She discussed and then recalibrated the product team's expectations to what a product marketer would actually do. They eventually hired well, but it was a reminder that even with veteran executives, surprising disconnects around product marketing still happen.

This misalignment in expectations is unfortunately not rare. Because the scope of product marketing isn't well understood, each team assumes it is the salvation for whatever gaps they feel most. In Roadster's case, the product team wanted to fill a gap around product adoption while the marketing team wanted to fill a gap to better leverage product in go-to-market.

This chapter attempts to level-set expectations for product marketing. It lays out a vision of what should be happening at major company stages. It should answer whether what you're expecting is reasonable and calibrate hiring for the right skills at the right time.

Early Stage: Ignition

This is a time of massive learning. It's when go-to-market fit gets discovered and frequent iterations are the norm. You're trying to find what ignites growth.

The task is to learn what consistently converts people into valuable customers. There should be a lot of testing of theories. Some can only be answered with time (such as best customers to retain).

Early go-to-market feels like hand-to-hand combat. Everything feels urgent and important. Companies at this stage aren't ready for bigger marketing budgets because they don't yet know who their best customers are. Ditto for the best ways to grow. As the product go-to-market strategist, the product marketer should be at the forefront of interpreting market signal and translating it into go-to-market actions.

There will be a lot of back and forth between anyone talking directly with customers—design, product, sales, customer success, and customers themselves—and product marketing. At regular, frequent intervals, the team discusses what is learned and priorities going forward. Product marketing then helps spearhead go-to-market response on all things product-meets-customer.

If you're a B2B company, a sales playbook should be taking shape. Test if you have repeatable steps. Messaging is likewise getting refined based on marketing activities or content that gets strong responses. Who your best evangelists are and what forms of evangelism have impact is also being discovered. Who are the best influencers? What communities spread the word?

This is also when it's important to lead with a point of view as much as what's in your product. The world must understand why your approach matters, why now, and why it's better than what the world already does. The ability to craft story using elements beyond just technology and product is crucial to get to the next stage.

This diverse range of skills requires a product marketer with more experience, typically at the director level. Once the product marketing foundation is in place, additional marketing resources can be added to amplify messages and drive campaigns.

Companies graduate out of this stage when go-to-market feels repeatable for two to three quarters and the company has more confidence in the customer segments where it succeeds.

Growth Stage: Rapid Rise

This is the fuel-on-a-fire stage because you have a baseline of what works. Success feels (mostly) repeatable. The challenge becomes, what got you here isn't necessarily what will keep you growing. Look to product marketers as enablers of business strategy, especially for newer growth opportunities.

Advanced PMM

Competitive forces drive a lot. Bigger companies might add functionality that competes in your space. Newer upstarts might be doing something different that takes advantage of a new technology or offers a similar service for less money.

For product marketers, helping the company own a position requires an elevated story and steady cadence of activity. It must expand people's notions of your category and your product's position in it. You also should be actively cultivating whatever drives evangelism—influencers, analysts, pundits, raving fans—and activate them with what they need to tell your story.

Product also gets more expansive, adding more functionality or entirely new products. Product marketing works with product to ensure additions matter to the market. The *why* for everything needs to be integrated into a bigger story that go-to-market teams can talk about.

Beware of stories getting too complex or too far ahead of what the market can bear. Product marketing also brings important market realities back to product teams and spearheads how to package messages and releases.

Product marketers also put in place processes that keep product and go-to-market teams connected.

Companies may also be adding additional go-to-market models. For example, a company that has been focused on direct sales up to this point may now introduce product-led growth (or vice versa). Product marketing's role supporting those different models shifts their scope.

Growth is of outsized importance at this stage. You emerge from it when the business has a pattern of many quarters in a row of strong, continued growth and revenue matching expectations of a company of your stage/size.

Mature Companies: Peak Burn

The line between growth and mature companies is fuzzy. It can be revenue, size of company, or sometimes simply how long a company has been around. It's very situationally dependent. The challenge of this stage is to maintain sustained growth.

Ironically, a mature technology company can have immature product marketing. This is evident if everyone is doing their own variation on a product marketing theme.

This is a time to clarify purpose and scope for the role. As companies mature, product marketing's impact is amplified— meaning it's game changing when done well and debilitating when it isn't.

Take this example from a meeting where sales, marketing, and product gathered to coordinate the company's response to their biggest competitor. Sales identified the most urgent issues:

- By the time sales was in an account, the competitor's executives had already established relationships with key stakeholders.
- The competitor invited a product bake-off on a part of the product where they knew they would win.
- The competitor's product approach felt newer; that gave it credibility.
- The mature company had been in market for much longer and people had preconceived ideas of what their product could and couldn't do.

Although these were all urgent sales issues, the immediate response was co-ordained by product marketing:

- **Sales qualification needed improvement.** The competition was disciplined at finding customers where they knew they could win. Product marketing worked with sales to revisit customer segments and qualifying criteria.
- **A positioning frame needed to be set before diving into features.** Controlling the conversation of why an approach is different and better moves the conversation away from features. Product marketing created a stronger story that positioned the product.

Advanced PMM

- **The product's advantages weren't memorable.** Claims without enough evidence or proof don't stick. Product marketing created more customer success stories and data to support the benefits and presented them in more memorable graphics and collateral.
- **Competitive repositioning tools were needed.** Sales needed to know how to better respond to what the competitor was saying and doing. Product marketing gave the team competitive battle cards with "kill points" on how to respond and win. It let sales sound smart at a moment's notice. They also worked with product to make sure future versions matched or beat the competition in key areas where the competition was beating them.

This is typical product marketing work at a mature company. They stitch together responses across product and go-to-market that are critical in highly competitive markets.

Beyond urgent market situations, product marketing focuses on longer-term market goals. For example, if a company wants to change the revenue mix over the next 18 months, product marketing sets in motion the partnerships, positioning, packaging, and sales incentives making that goal possible.

Product marketing also makes sure a foundation is in place for strong evangelism by others—including all sales enablement but also the larger influencer ecosystem that dominates digital channels. For mature companies, there is also a lot more invisible influence occurring. Working with marketing teams to understand market sentiment is important.

This is a phase when marketing work is not evenly distributed across all products or lines of business. Older products may need technical upgrading but little marketing support. A new product might require a lot more marketing investment to become a reasonable business. A product suite important for the company's future growth might require a lot of extra product marketing work. Product marketing should be intentionally applied to the areas of greatest market need. For mature companies, not all products are equal.

Product marketing can be a truly powerful business enabler at this stage. There are enough resources to be good at both the urgent and longer-term go-to-market work. It's also why clear definitions of what the company expects from the role are so important.

Adjust Scope to Stage

The gap between the aspirations of product marketing and reality often leads to needless disappointment. Take the time to keep the organization's expectations of the role aligned among *all* teams, not just the ones product marketing works with frequently.

Throughout all the stages of a company, product marketing's core responsibilities stay relatively stable. It's the range, focus, and complexity of how they apply them that changes as companies mature.

The key at any stage is to be unafraid to adapt the scope of the role and to communicate it clearly when you do. It's not easy, but it's what makes impactful product marketing possible.

Chapter 30

Mature Company Inflection Points

When to Lean into Product Marketing

Advanced PMM

A 90-year-old insurance company began thinking of their customer data as a product to power better insurance decisions, not just for themselves but for other companies.

A 20-year-old software company with dozens of products learned CIOs didn't recognize their brand outside of their flagship products so began shifting to integrated product suites.

A 15-year-old company was launching an Amazon Prime–like subscription membership as a completely new way to access all of their services.

A 10-year-old company with an English-language product already used in most countries of the world wanted to expand its international business to grow.

These real-life scenarios show inflection points driven by changes in product that likewise require shifts in go-to-market to succeed. But often, of equal or greater importance is the unlearning—both for internal teams and for how a market thinks about a company.

Major business initiatives like these are massively coordinated at the company level with full support of executive leaders. But any time a major product initiative requires the market to rethink a company, product marketing is a powerful catalyst in the transformation process.

These are times to lean into product marketing and leverage the unique strengths of the role.

"Traditional" Company Becomes a Tech-First Company

This more than 100 year old company is famous for their construction equipment yet employs more software than mechanical engineers!

It's part of the company's move toward precision construction, where their equipment acts as a mobile sensing unit continuously streaming data to the cloud so their customers can do their jobs smarter.

The company has corporate marketing teams whose job it is to make everything feel and look on brand and communicate the company's bigger vision. They also have internal process groups whose purpose is to drive efficiency through standard processes and make sure they get used.

When they made the switch to become a tech forward company, they completely reorganized their product organization and how it worked. Empowered product teams were more agile and producing software products the company wasn't used to marketing.

That's when they started rethinking the role of product marketing teams. New product directions needed to get better connected with the company's more traditional go-to-market approach. Their first "customers" were internal ones: marketing, process, dealer, and field sales teams.

Product information needed to go beyond product comparison videos or website details. It needed to start consistently including context around the role of data and technology in the quality of their customers' lives. They created proof of why the new tech was better, such as a time study using the sensor guidance systems versus how much longer it took doing the job without them. For early adopters, specifics matter—vision and brand aren't enough.

Advanced PMM

The work required identifying important influencers—like different generations of customers—and not just early users. Shifts in go-to-market also meant trying new things optimized for early adopters that felt outside the company's normal marketing practices—like a short Valentine's Day video on social which was loved by tech-forward followers but not as loved by corporate marketing.

This tension—both internal and in the market—is natural and necessary as markets and traditional companies get pushed. Product marketers in this situation must remain steadfast in nudging the market and internal groups forward. That consistency, as much as any individual activity, is what eventually creates desired change.

A Single-Product Company Becomes a Multi-Product Company

This inflection point happens for every growing tech company. Some reach it sooner than others. Some do it sooner than they should. It's inevitable if you're succeeding.

The challenge lies in organizational and market muscle memory. Sales teams are more comfortable selling what they already know. Customers don't rethink their notions of a company unless continually pushed.

Getting past all this natural inertia means product marketers must resist the temptation to fall into *mo' product, mo' better* mode. Customers need to understand why the approach for a suite of products makes it a better way to solve their problems. Establishing the story framing the *why* is important before diving into any product specifics.

In the best case, people focus on the value of a solution and not which product does what. Remember product dividing lines often reflect a company's organizational structure or maturity, not necessarily how customers want to experience value.

Just as important, don't assume the same GTM motion that worked for one product works the same for those that follow. Not only is competition different but buying centers and market dynamics may vary. Product marketing needs to understand where there is or isn't overlap and direct the company and its go-to-market accordingly. This is especially true in the case of an acquired product.

Because product marketers typically have visibility into major new products in the future, it's important they spearhead preparing the company *and the market* for this transition.

It means a lot of tools and training is necessary for go-to-market teams to succeed. The influencer network must be activated. Thought leadership around why an approach matters needs to be discoverable based on how people search. And of course, how to evolve the company's brand is critically important. When go-to-market is proactive and coordinated on all these dimensions, it has outsized impact. Product marketing is key in defining the strategy and guiding its execution.

Moving from Product to Solution, Service, or Customer-Centric

Back when Microsoft Office was new, you could choose from two different editions—Standard or Pro. The differences were based on which products were included. You decided the edition based on the products you wanted.

In the subscription economy of today, Microsoft has shifted away from products as the anchor of value. They focus, instead, on customer segments: Microsoft 365 for home or Microsoft 365 for business. Product and cloud-based services move fluidly across the offerings and change whenever they need to. Customers subscribe to the service for its total value, not for any individual product, although enough known products are

Advanced PMM

included to create a baseline for why customers should value the service.

This is a prime example of service, solution, or customer-centric go-to-market versus product-centric. Value is constructed around a market segment and providing ongoing value to those customers. The value extends far beyond the products being bundled together.

This takes many forms, from subscription services to licenses for unlimited access. The later as a tactic is particularly well suited to companies whose product portfolio is large. Microsoft's multiyear enterprise licenses are what help many of their new products grow adoption so quickly—it's already included in what companies paid for, so it gets used.

This work is fundamentally about how to frame and package value to improve adoption in desired markets—squarely the product marketing sweet spot.

International Expansion

Account-based marketing (ABM) was a well-established term in the United States, so much so that the US-based mothership Demandbase was moving on to its next big message. But the European market lagged the US. The head of European sales still relied on the company's older ABM messaging to connect with prospective customers. Anything different was too far ahead of his market.

It's a great example of markets in different geographies being distinct. Go-to-market tactics and messaging must be tailored to the specific needs of local markets. Product localization and internationalization work aside, it's why international growth is always a big deal for any company.

This is where local country go-to-market teams are essential. It usually involves in-country field and marketing roles that know how to take what they get from HQ as a starting point, then make it work for their geography.

Advanced PMM

Product marketing can be a powerful translation layer between in-country go-to-market specialists and the work happening at HQ. Product marketers focused on international tend to be overlays across multiple products or services. They represent and prioritize local market needs to product and go-to-market teams so they, in turn, make decisions from a more global perspective from the start.

They also communicate nonobvious marketing challenges: a feature name that doesn't translate well in a particular language, differences in technology accessibility, or product behavior that doesn't culturally translate in other parts of the world.

Mature companies are in the unique position of being able to apply more product marketing resources in advance of hitting company inflection points. Doing so can amplify go-to-market success. But just as importantly, it's a way to keep internal teams affected by product go-to-market better aligned.

Regardless of your company's maturity, inflection points should be seized as an opportunity to leverage the unique power of product marketing.

Advanced PMM

Conclusion

What You Can Do Right Now

My friend Grady Karp was the principal product manager for the inaugural Amazon Echo. I was on the edge of my seat as he told me about his adventures creating a voice-only user interface and how Jeff Bezos basically acted as their early product marketer.

Most of us aren't lucky enough to work on a category-defining product with a once-in-a-generation CEO as the go-to-market nudge. We work on interesting products, on pretty good teams. The idea of going from good to great is intimidating and hard.

The model of product marketing I prescribe in this book is the role at its most powerful and capable. Not every person or organization will get there, but it is worth striving toward.

I hope the examples used throughout the book inspire you for what success can look like while also confirming some of the inevitable, hard realities that are simply part of the journey.

If all you do is get better at applying the four fundamentals, you'll already be ahead of most.

Four Fundamentals of Product Marketing

Fundamental 1: Ambassador: Connect Customer and Market Insights
Fundamental 2: Strategist: Direct Your Product's Go-to-Market
Fundamental 3: Storyteller: Shape How the World Thinks About Your Product
Fundamental 4: Evangelist: Enable Others to Tell the Story

But as you strive toward the ideal, the following is a list of things anyone can do with any product marketer in any company structure and still make a difference:

- **Ask for the market or customer point-of-view in product and go-to-market meetings.** Don't wait for this to get volunteered. If you're building, marketing, or selling something, there is a customer or market point-of-view that will enrich how you approach the task at hand. Ask product marketers for this every chance you get.
- **Learn what's working well in marketing.** If marketing does a particular activity that has outsized performance relative to everything else, there might be an important market signal in it—around audience, messages, or what features are most compelling. Not all things are equal when they get to the market.
- **Share more stories.** Stories aren't just powerful for marketing purposes, they're equally powerful for bringing ideas to life for internal teams. Don't roll everything up into just data factoids. Tell stories about real people, doing real things, and the real impact your product is having.
- **Revisit messaging.** It's rare that messaging can't be improved and made more customer-centric in some way. Use the CAST guidelines to assess where yours can be better: (1) Is what you do **clear versus comprehensive**? (2) Is the language meaningful and **authentic**? (3) Is what makes you different **simple** enough to be understood?

(4) Have you **tested** it in the context customers see it? And use the examples in this book for inspiration.

- **Use the Product Go-to-Market Canvas to create alignment between go-to-market teams and product.** Every size organization benefits when they connect customer realities with what's planned in marketing, sales, *and* product. This shapes product go-to-market strategies. They, in turn, make any marketing activity meaningful.

- **Use the Messaging Canvas to improve what marketing and sales teams say.** Especially if you're in an organization where the market-facing teams are large and functionally separated from product marketing, this is one of the most impactful ways to ensure product-IQ gets translated to how the rest of the company markets and sells the product.

- **Use a Release Scale to develop shared expectations between product and marketing.** Shared vocabulary and expectation management is half the battle with high velocity teams. Get everyone on the same page, and don't overthink it. A lot of working better together is about educating people on what's possible and inviting productive debate where teams see things differently.

- **Use agile marketing practices.** Have product marketers lead weekly reviews (scrums) discussing marketing activity to prioritize and improve. Ensure the latest market learning informs what gets done in each sprint.

Every company can improve product marketing. And every tech businesses does better when product marketing is better. The journey begins by rethinking assumptions about the role and having the courage to expect more.

I wish you fair winds and good luck in your journey.

Appendix: Marketing Terms Explained

Account Based Marketing.   Aligns marketing and sales by examining actions taken by multiple people at the same account (company) to determine what the next best action is to convert an account into a customer. It uses behavior (someone looked at this web page) and intent data (they then looked at a competitor's website) of everyone in a company regardless of role to determine if they are a good lead. It typically takes more sales touches than marketing touches to get someone to the end of a B2B funnel, but the approach is most cost-effective when it's coordinated as it is by design with ABM.

Most used in B2B companies with target account lists.

Affiliate marketing.   A type of performance-based marketing where there is a predetermined economic arrangement (commission) for marketing or referring a customer to a product.

Most used with consumer-facing goods and when someone else owns a distribution or referral platform that has a lot of your desired customers (e.g. referral website).

Analyst relations.   The collective work with analyst firms, like Gartner, Forrester, or IDC (or boutique specialist firms), who have dedicated industry analysts and produce evaluations of most major technology market categories.

Most used in B2B markets to validate products relative to key competitors. They can be particularly helpful for companies wanting third-party validation of their value in a category. Some companies will only buy software from companies reviewed or rated by analyst firms.

Brand.   Comprehensively defines your relationship with your customer—centered on core attributes that the best companies choose very early in their journey. It is distinct from logo, although logo is the iconic, visual representation of your brand.

Most used Everyone should be intentional about their brand. 90% of consumers already know the brand they'd like to buy before they begin active consideration (Google).

Call to Action (CTA).   In most marketing activities, there is always some kind of CTA, the next step a marketer wants someone to take.

Most used Nearly every marketing activity has one; it's how marketing knows it's creating the desired outcome.

Channel marketing.   The blanket term used to describe different channels—not direct via a salesforce or to consumers—you sell a product through. Retail and online are channels. Distributors are channels. Original equipment manufacturers are channels. Major consulting firms could be channels.

Most used Every product must find its own path through the best channels for that product. You learn the most when you go direct to customers because you have immediate access to all the data and market signals, but channels are effective if partners own existing relationships with your desired customers.

Community marketing.   Creating a group of people who share affinity over a problem (environmental sustainability), product (like a user group), or role (like a group of CMOs or VCs) to network, learn, and support one another over a topic and/or a product. It's also a way to create a starting point for evangelism.

Most used This is commonly used at companies with big enough customer bases that a community helps scale product or category knowledge with customers instead of a company leading discussions directly.

Content marketing.   Points of view with truly helpful information. *It's a sales-free zone,* which is how it's different from traditional marketing. It's the modern approach to establish expertise and thought leadership. When COVID first broke out, the person with outsized global reach, impacting policy decisions before official institutions were ready, was a VP of Growth, not an epidemiologist or scientist. He simply compiled great global data and explained what was happening in a way anyone could understand.

Most used This is considered a marketing standard like having a website. When people search, this is how you cast a broader net to be found. NerdWallet's entire business was built around a content engine. HubSpot popularized it as the foundation for inbound marketing.

Corporate marketing.   Marketing at the company level, versus the product level. Depending on your stage, go-to-market, and strength of products, this will vary in importance and relative weighting in the overall marketing mix.

Most used when you're ready for the company to be seen as a separate unifying entity beyond product.

Crowdfunding Campaign.   Kickstarter, Indiegogo or any other of many platforms. They are a place where you can trial market interest and initial positioning and messaging. It's particularly good for understanding who early adopters might be, getting some feedback on pricing, and closing the "say vs. do" gap. It's a place where people can translate their interest into action and vote with their dollars.

Most used by physical products that might still be in very early stages of development, even still just in the idea phase.

Customer funnel.   Each step of the customer's journey from awareness to purchase. Dave McClure's pirate metrics (AARRR) popularized acquisition, activation, retention, revenue, referral. Marketers typically refer to activities as they affect different parts of the funnel: TOFU = top of funnel (awareness, acquisition, activation), MOFU = middle of funnel (consideration, evaluation, referral), BOFU = bottom of funnel (decision, purchase, retention, advocacy).

Most used Every business needs to have this instrumented and measured. You can't market effectively without visibility into how this works for your company.

Demand Generation/Lead Generation/Pipeline Generation. The collection of activities that drive awareness and interest to eventually become qualified leads that convert into a pipeline for sales.

Most used Every B2B business has demand generation specialists whose purview can be broad (events, digital campaigns) and whose success is measured by the qualified leads their marketing programs generate.

Demo.   Video or in-person demonstration of your product's highlights and typical use cases. This should never just be a feature walk-through. Best if done through true use cases and highlighting most important features that reaffirm your position, not showing off everything.

Most used Should be used in some way by every company.

Direct-response marketing.   Any marketing trying to elicit a direct response to a call-to-action, such as click on this blog post, attend this webinar, or here's a free gift we've mailed to your desk or door. This is in contrast to marketing focused on generating awareness and affinity for a brand, for example, a campaign promoting content of recent award winners.

Most used in demand generation activities where the goal is to get someone into a customer funnel, not just create awareness or nurture a relationship.

Event marketing.   Either your own event (like Dreamforce for Salesforce) or attending others' events that have established reputation in your industry (like RSA for cybersecurity), it's a way to rapidly gain exposure to a particular target audience with the benefit of physical energy and serendipitous exposure. It's a great place to partner with others and stand out among your competition.

Most used by companies who have specific segments they target. If you do them, stand out. Their biggest value is often trying to get a spot on a stage in some way and getting direct access to attendee lists.

Influencer marketing/social influencers.   Getting endorsements or product placements from influencers with knowledge of your space that also have large amounts of followers.

Most used Essential for direct-to-customer brands, but every company should do this as influence is increasingly decentralized, hard to track but effective.

Jobs to Be Done.   A framework for product development from Tony Ulwick, popularized in his *Jobs to Be Done* book. It frames the customer's specific "job" they are trying to get done and their true motivations for using a product to complete the job. It's intended to help product teams uncover underlying goals that motivate a customer.

Most used by some modern product teams to frame their customer discovery work.

Partner marketing.   You are judged by the company you keep. Creating industry validation and increasing sales channel or customer access through partners who already hold the desired relationships you seek. This can range from a marketing partnership (e.g. do events or campaigns together) to revenue relationships. Also see channel marketing.

Most used Should be used in some way by most companies at the right time(s).

Performance marketing.   Paid advertising, in which one pays only when there are measurable results. Paid advertising like this can very quickly eat through a budget, so frequent monitoring is necessary.

Most used as a way to broaden digital awareness, elevate a campaign, or promote a specific content asset intended to drive marketing leads.

Press release.   A standard format designed for the who, what, when, where, why, and how for how journalists write stories. They want "just the facts" so they can write their stories with their point of view. The easier you make the facts to find, the better. You typically minimize promotional language in a press release, as that makes it feel less credible to anyone in the press. Press releases get "dropped on the wire" where they can get seen and picked up by many news services.

Most used at a minimum when there is news to announce (Company X just got $50M in funding) or show momentum (signed Walmart as a customer) but also for link juice to help improve search engine discovery.

Product-led growth (PLG).   Go-to-market strategy that relies on the product to acquire, activate, or retain customers. It is seen as one of the more cost-effective ways to create a customer base, in particular to engender organic growth and evangelism. It's particularly popular for developer tools, as it's always presumed developers like to try something before they believe it will work for them, as well as consumer-facing companies. In B2B companies, it can mean using product data to drive sales actions. The idea with PLG is people like the product so much, they use it with others, which grows the number of users organically or vastly simplifies the sales process.

Most used in businesses with a direct relationship to their customers but increasingly, part of a blended go-to-market approach for enterprise software companies.

Public/press/media relations (PR).   Press is third-party validation that shows the world you're worthy of being in the news. It can be particularly helpful for recruiting, customers, and short-term spikes in awareness and website traffic. It is most powerful when you can build and maintain ongoing relationships with media so you're consulted as an expert in your field. That way you can get pulled into stories versus just waiting for stories to be written about your company. Note that staff at all publications are spread thin and get hundreds, in some cases thousands of requests for coverage a day. Relationships and really thinking through why you're newsworthy (not just that you think you are) is critical.

Most used by companies seeking validation or promotion of a newsworthy milestone.

Sales enablement.   The collection of activities that provides sales with the background materials, content, tools, training, and processes that let them sell more effectively.

Most used by every organization that has a direct sales force. Also used in channel marketing to arm channel partners with materials to sell successfully.

Search engine optimization/marketing (SEO/SEM).   Using specific techniques to improve the overall ranking and discoverability of your website and content in search engine results. It includes finding related terms and searches you want your product associated with. SEM is the paid aspect of this. It's the advertising in support of key search terms so content, products, services, or companies get discovered when a customer is searching to solve a problem.

Most used Every website and content strategy should be informed by some SEO work to ensure a website is discoverable and ranks high enough to be found by customers. This may or may not include SEM, which tends to be something companies can do when their digital marketing budgets or presence grows.

Social media marketing.   Using social media platforms as the way to promote your product, company, or employees. This can be your own company area, promoted posts, and/or also include influencer or content marketing. Great for creating or deepening brand loyalty, amplifying something important, and engaging in two-way conversations with customers.

Most used It's a great channel for activating evangelism in a more organic way that's not just about the product or company.

Sponsorships.   Financial or in-kind support to have your company associated with an event, venue, or organization. It's a way of buying awareness through organizations or events that already have an existing relationship with your target audiences.

Most used for brand awareness and "validation by association." It's about associating your brand with an event or customer segment you desire.

Technical evangelism.   Common in developer marketing. These more technical specialists are able to "talk tech" with any developer to specifically advocate for a technology or product's use in technical (code, architecture) use cases. These people are often developers or engineers themselves, but at the very least are users of the product.

Most used anytime you're marketing a technical product, service, or APIs to developers.

Traditional advertising.   Typically refers to outdoor, radio, TV—non-digital—forms of advertising.

Most used You need either a certain amount of resource or need for specific reach for this to make sense, but if you're targeting specific geographies, industries, or demographics, it can be a great way to generate awareness.

Word of mouth (WOM) marketing.   Actively   influencing or encouraging people to publicly talk or write about your product or service. Sometimes used interchangeably with organic growth or social influence, it provides evidence of genuine product love. Examples include: customer reviews of products on comparison sites or postings in developer forums or communication channels.

Most used People look for positive WOM before making buying decisions. The more technical or social (e.g. younger demographics) the audience, the more they prefer these channels for learning.

Acknowledgments

Writing this book scared me. I knew it was going to be hard because writing anything worth reading is. It's given me a deep appreciation for creators for whom an audience is the final arbiter of worth. I owe a particular debt of gratitude to Malcolm Gladwell's Masterclass, which inspired me to try new things as a writer. To all the other authors I read, directors I watched, and composers whose music inspired me, thank you. I consumed a lot of your work to help fuel mine.

To Chris Jones—the love of my life and my partner in all things—and Anya and Taryn—you could not have been more supportive. For your understanding when I was chained to my chair and your patience when I wrote, I can't express enough gratitude or love. To my family—Birthe, Duane, Pam, Lance, Susan, and Mom and Dad (RIP)—I am me because of you.

Any advice I'm dispensing in this book has been garnered from the incredible experiences I've had working with great people at great companies. There are too many to mention all by name, but if you've been part of my learning journey, know that I appreciate every lesson—the bad as much as the good. My early years at Microsoft, Netscape, and Loudcloud were particularly formative, so to my colleagues and friends from those companies (Sarah Leary, Michael and Kathleen Hebert, John Wood, Blue/Peter Pathe, Jeff Vierling, Eric Levine, Lynn Carpenter Schumann, Eric Byunn, Eric Hahn, Bob Lisbonne, Ben Horowitz, Marc Andreessen, Jerrell Jimerson, Brian Grey, Tim Howes, and Mike Homer [RIP])—you've left an indelible impression and shaped the professional I've become. Team

Pocket—Nate Weiner, Nikki Will, Matt Koidin, and Jonathan Bruck—thank you for letting me ride along on your journey.

If this book is worthy of reading, it is thanks to a small but mighty group of reviewers who pushed me anywhere my ideas could be stronger. They were specific in their feedback, candid, encouraging, and extremely generous with their time: Gabi Bufrem, Scott Guidoboni, Kenaz Kwa, Tony Liu, Tatyana Mamut, Kevin McNamara, Jay Miller (twice!), Jim Morris, Rachel Quon, and Matt Stammers. I cannot thank all of you enough. This book is so much better for all of your input and care.

I also knew I needed to become a better writer through this process and had the gift of Leslie Hobbs as my coach. She's been with me since page one, reviewing and editing every draft over more than 18 hard months. She encouraged me when I needed it most and made me believe I could do this. Leslie— I'm beyond grateful.

Lauren Hart designed the book's graphics and was a wonderful thought partner to improve not just graphics but clarity.

This book was made possible by every part of the Costanoa Ventures family. My work with Costanoa's startups pressure tested my frameworks, illuminated product go-to-market challenges, and made me better at everything I teach. Thanks for letting me be part of your journeys. Greg Sands, I have such deep gratitude for you making the space for all of me and this book at Costanoa. To my operating partner team—Michelle McHargue, Jim Wilson, Bettye Watkins, Taylor Bernal, Katy Wiley, and Rachel Quon (honorary)—thank you for creating the space to help me get this book done. To the rest of the investing team—Amy Cheetham, John Cowgill, Tony Liu, Mckenzie Parks, and Mark Selcow—I appreciate your trust in my expertise. Nancy Katz—special thanks for your copyediting. And Pamela Magie, Chonlana Jarawiwat, and Mike Albang, thanks for making Costanoa such a great place to work.

But this book's origin story and raison d'etre lies with SVPG (Silicon Valley Product Group) and Marty Cagan. You have *INSPIRED* and *EMPOWERED* me by your example and are the reason why I'm surrounded by a world-class group of partners I count myself immeasurably lucky to be one of. I feel immense gratitude for all of your partnership. Lea Hickman, Chris Jones, Christian Iodine, Jon Moore, and Marty, thank you for your kind counsel, in-depth and incisive feedback on my drafts, and encouragement throughout this process. You challenged me in the best possible ways. The book and I am better for it.

About the Author

Martina Lauchengco is the product marketing partner at SVPG (Silicon Valley Product Group) and a partner at Costanoa Ventures, a boutique early-stage venture capital firm. Martina worked at Microsoft and Netscape, two of the most formative tech companies of all time, before advising startups and Fortune 500 companies. She is a lecturer in the engineering graduate program at UC Berkeley. Martina spends most of her time working with early-stage startups doing exactly what she writes about in this book. She is also an ardent advocate on the importance of diversity, equity, and inclusion. Her writing has been featured in *TechCrunch* and *VentureBeat.* This is her first book.

Martina received her BA and MA from Stanford University, where she was a collegiate cycling national champion. She still loves all things bike and run and spends winters snowboarding and cross-country skiing. She lives in San Francisco with her husband, two children, and collection of OV Episode IV, V and VI action figures.

Twitter: @mavinmartina
LinkedIn: martinalauchengco
Martinalauchengco.com
Lovedthebook.com

To learn more about how great products are built or attend a workshop on how to rethink product marketing visit www.svpg.com.

To learn more about how Costanoa Ventures invests in Seed and Series A companies that change how business gets done in enterprise SaaS, data infrastructure, and financial tech, visit www.costanoavc.com.

CostanoaVentures

Index